T0222633

Materials, Integration and Technology for Monolithic Instruments

MATERIALS RESEARCH SOCIETY
SYMPOSIUM PROCEEDINGS VOLUME 869

Materials, Integration and Technology for Monolithic Instruments

Symposium held March 29–30, 2005, San Francisco, California, U.S.A.

EDITORS:

Jeremy A. Theil
Lumileds Lighting, LLC
San Jose, California, U.S.A.

Markus Böhm
University of Siegen
Siegen, Germany

Donald S. Gardner
Intel Corporation
Santa Clara, California, U.S.A.

Travis Blalock
University of Virginia
Charlottesville, Virginia, U.S.A.

Materials Research Society
Warrendale, Pennsylvania

CAMBRIDGE UNIVERSITY PRESS
Cambridge, New York, Melbourne, Madrid, Cape Town,
Singapore, São Paulo, Delhi, Mexico City

Cambridge University Press
32 Avenue of the Americas, New York NY 10013-2473, USA

Published in the United States of America by Cambridge University Press, New York

www.cambridge.org
Information on this title: www.cambridge.org/9781107408951

Materials Research Society
506 Keystone Drive, Warrendale, PA 15086
http://www.mrs.org

First published 2005
First paperback edition 2013

Single article reprints from this publication are available through
University Microfilms Inc., 300 North Zeeb Road, Ann Arbor, MI 48106

CODEN: MRSPDH

ISBN 978-1-107-4089-51 Paperback

CONTENTS

*Invited Paper

CHEMICAL AND BIOLOGICAL
SENSING SYSTEMS

*Invited Paper

FUNCTIONAL OXIDES AND OTHER MATERIALS
FOR MONOLITHIC INSTRUMENT INTEGRATION

*Invited Paper

PREFACE

This volume, containing papers from Symposium D, "Materials, Integration and Technology for Monolithic Instruments," held March 29–30 at the 2005 MRS Spring Meeting in San Francisco, California, is the first to bring together various facets of research into developing an exciting new class of highly integrated devices, called Monolithic Instruments. Monolithic Instruments are miniaturized systems which interact with their physical environment in ways traditional integrated circuits cannot, by combining conventional integrated circuits with novel solid-state components. These systems are enabled by utilizing the extremely precise manufacturing platform that is the integrated circuit wafer fabrication facility.

The semiconductor industry has transformed society at the turn of the Millenium by bringing out ever increasing computing power for ever lower cost. This has largely been achieved by the extremely high degree of precision that the semiconductor industry has achieved. The ever increasing presence of integrated circuits in products has enabled electronics to take over most system functions in many products, from cameras to televisions, from spectrometers to gas chromatographs. The functionality now handled by integrated circuits, includes the analog/digital interface, signal processing and conditioning, signal analysis, all the way through to the user interface. The only function that is outside the realm of integrated circuit capabilities is that of transducer/actuator fabrication. However this is changing as researchers learn how to successfully integrate new transducers and actuators with integrated circuits. The Monolithic Instrument concept is quite powerful in that it enables vast cost and size reductions through the use of integrated circuit manufacturing platforms to build such instruments. For example, DNA microarray systems can be fabricated for a few dollars that perform similar functions of advanced DNA array/scanner systems that cost tens of thousands of dollars.

The papers presented in this volume are a subset of what is possible in this field. The first section examines advanced image sensor concepts by forming the photodetector above the standard CMOS interconnect. It discusses not only how such devices may be made but the performance of such devices, and the challenges in designing circuits incorporating them. The next section includes other optoelectronic element integration, including the critical components for constructing miniaturized spectrometers. The third section covers the development of liquid chemical sensing systems including issues with successfully integrating such sensors with CMOS circuitry, and how to create sensitive compact DNA sensing systems. One of the most readily integrateable class of materials are oxide films. There are several papers on oxides that form the core for various sensor and optoelectronic systems. We hope in the future to have contributions from other applications in such fields as low-cost gas sensing (electronic noses), and compact emissive microdisplays to name a few.

The integrated circuit field is now on the cusp of a new level of integration that can enable an entirely new class of products, Monolithic Instruments. We hope that you learn and enjoy the papers in this volume and become as excited as we are for the new possibilities this field presents.

Jeremy A. Theil
Markus Böhm
Donald S. Gardner
Travis Blalock

June 2005

MATERIALS RESEARCH SOCIETY SYMPOSIUM PROCEEDINGS

MATERIALS RESEARCH SOCIETY SYMPOSIUM PROCEEDINGS

Prior Materials Research Society Symposium Proceedings available by contacting Materials Research Society

Imaging Sensing Systems

Vertical integration of hydrogenated amorphous silicon devices on CMOS circuits

N. Wyrsch[1], C. Miazza[1], C. Ballif[1], A. Shah[1], N. Blanc[2], R. Kaufmann[2], F. Lustenberger[2], P. Jarron[3],
[1] Institut de Microtechnique, Université de Neuchâtel, Breguet 2, CH-2000 Neuchâtel, Switzerland,
[2] CSEM SA, Badenerstrasse 569, P.O. Box, CH-8048 Zürich, Switzerland,
[3] CERN, CH-1211 Genève 23, Switzerland.

ABSTRACT

Monolithic integration of sensing devices usually requires sharing the CMOS chip floor space between sensors and their readout electronics. Vertical integration of the sensor on top of the electronics allows one to have the full chip area dedicated to sensing. For light detection, the deposition of hydrogenated amorphous silicon (a-Si:H) photodiodes on top of CMOS readout circuits offers several advantages compared to standard CMOS imagers. The issues regarding the design of a-Si:H photodiodes, their integration and the influence of the CMOS chip design (i.e. its surface morphology) on a-Si:H diode performance are discussed. Examples of TFA sensors for vision and particle detection are also presented.

INTRODUCTION

Active pixel sensors (APS) in CMOS technology have recently gained a lot of interest. Many functionalities can be implemented at the pixel level, ranging from basic charge integration or amplification to pre-processing of the data. However, the fact that the pixel readout electronics shares the die area with the sensor element is an important factor limiting the sensitivity (the sensor area being limited) and leads also to "dead areas" which are unacceptable for certain applications. The introduction of more advanced technologies (with smaller feature sizes) renders the problem more acute because the sensor area is further reduced and it introduces difficulties for the coupling of the sensor through the various metal layers of the chip.

Vertical integration of hydrogenated amorphous silicon (a-Si:H) sensors on top of readout electronics is a promising solution to this problem. This concept has been introduced successfully for several applications, especially for vision sensors with high sensitivity [1, 2] or high dynamic range [3].This integration concept is known as thin-film on ASIC (TFA), thin-film on CMOS (TFC), above IC (integrated circuit) or elevated diode technology (in the cases where a diode is used). The pioneering work of the University of Siegen in Germany [4] on this concept has attracted world-wide a large interest in this technology, for imaging [5], color detection [6], but also for other applications, such as infrared light vision [7] and particle detection [8, 9]. MEMS (micro electro-mechanical systems) or BioMEMS (biological MEMS) are further possible target applications for TFA technology. Even though vertical integration may basically involve various types of materials and circuits, we will restrict ourselves here to a discussion of a-Si:H based devices on CMOS circuits. The typical structure of such a device is presented in Fig. 1.

For light detection, TFA technology offers several advantages compared to c-Si technology with embedded photodiodes:

(a) Maximization of sensitivity, since the entire chip area may be dedicated to light collection. The (geometrical) fill-factor may be close to 100%, i.e. 100% of the chip may be dedicated to active sensor area. For smaller feature size (advanced CMOS technologies), TFA technology may be the best option for high sensitivity [10].

(b) Clear separation between optimization of the photodiode and design of the CMOS circuit; this reduces constraints in the placement of the pixel circuit elements.

(c) Large flexibility in the choice of the active material for the photodiode ; this allows for tuning/adjustment of the spectral sensitivity. Various thin-film semiconductors (a-Si:H, silicon-germanium or silicon-carbon alloys, micro-crystalline silicon, etc) can be chosen or combined to extend or restrict the spectral sensitivity.

(d) The possibility of a vertical integration of several photodiodes forming a multi-junction device with two or multiple terminals in order to detect and/or separate various portions of the electromagnetic spectrum. Combination with other elements, such as a scintillating layer for X-ray to light conversion, is here a further option.

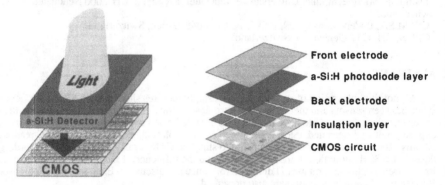

Figure 1. Schematic view of an array of sensors in TFA technology. In most cases, the CMOS circuit passivation layer is used as the insulation layer. The top metal layer of the CMOS chip is either used as the back electrode of the a-Si:H diode layer or an additional metal layer is evaporated on top of the chip. The array is defined by the patterning of the back electrode.

In this context, a-Si:H offers two significant advantages: (a) low deposition temperature (around 200°C), this means that there is no problem for the direct deposition of a-Si:H on CMOS chips, and (b) a larger band gap (larger than that of crystalline silicon), a fact that has beneficial aspects for many of the applications mentioned above. a-Si:H offers appealing characteristics for other specific applications such as high resistance to radiation [11],mechanical properties close to those of c-Si, possibility to vary the optical band gap by alloying of silicon with materials having lower band gaps (Ge) or larger band gaps (C or O). This paper will review some of the aspects of TFA technology, with an emphasis on the processing of a-Si:H layers and diodes as developed at the University of Neuchâtel. The issues regarding the design of a-Si:H photodiodes will be detailed; specifically, the influence of the CMOS chip design/topology on the performances of the a-Si:H photodiodes will be discussed. We will thereby consider the case of applications in low-level light detection and particle detection and present results recently obtained in our laboratories.

a-Si:H LAYER AND DEVICE PROCESSING

Deposition process

a-Si:H material is an alloy of silicon with around 10% (atomic) hydrogen. The presence of hydrogen is necessary to passivate the dangling bonds resulting from the formation of an amorphous tissue. The irregular arrangement of the Si atoms has the consequence that not all covalent Si-Si bonds can be satisfied. The minimal amount of hydrogen to passivate most of the dangling bonds is below 1%. The hydrogen content of the material has a direct influence on the band gap and is mainly controlled by the deposition temperature.

a-Si:H layers are, in general, deposited at deposition temperatures around 200°C from the dissociation of silane; hydrogen is generally used for diluting silane. The possible range of temperatures extends between 100 and 350°C, depending on the desired material characteristics (band gap, mechanical stress). The processing temperature is therefore fully compatible with CMOS technology. However, if multiple layers are involved for the fabrication of thin-film devices, the process temperature of any given layer should never exceed the temperature of previously deposited ones.

Several deposition techniques are available for electronic quality a-Si:H: Plasma-enhanced chemical vapor deposition (PE-CVD) with plasma excitation at the standard radio-frequency, as used for Industrial processes (RF at 13.56 MHz) [12], in the very-high frequency range (VHF – between 50 and 150 MHz) [13]or in the microwave frequency range [14], hot-wire (HW) deposition (also known as catalytic CVD) [15]. All techniques cited above have been able to produce high-quality a-Si:H material and have been successfully used, in the laboratory, for a-Si:H based solar cells. However, each one of them offers specific advantages and disadvantages, as partly given hereunder.

For TFA applications, several factors should be discussed: Plasma discharge voltages, powder formation in the plasma, deposition rates obtained, as well as for the resulting layers: defect density, uniformity, mechanical stress and material stability. Plasma deposition over a sensitive CMOS circuit is basically a critical point, in view of the ion bombardment and the voltage spikes that can be created in the plasma. However, the deposition of device-quality a-Si:H layers has , in general, to be performed with "mild" plasma conditions, i.e. under conditions where the ion bombardment is "soft", i.e. only low energies and relatively low electric fields are involved. Both VHF PE-CVD and microwave plasmas are in this context "milder": they lead to lower electric fields in the plasma sheaths, lower than what is obtained with RF-PE CVD. HW deposition eliminates the presence of a plasma, but requires a very precise control of the temperature of the growing film in order to obtain optimal layer and device characteristics.

The use of plasma cleaning prior to the deposition of the a-Si:H photodiode is, however, often a critical step, since in this case a "naked" CMOS chip faces a (relatively "violent") plasma (as is often used for cleaning operations). The authors have experienced cases where this lead to a destruction of the CMOS chip.

For photosensitive devices, defect density is an important material characteristic. Even though all techniques mentioned here allow the deposition of low-defect material, the defect density strongly depends on the deposition conditions. High deposition rates imply higher plasma powers which lead to high ion bombardment and larger defect density in the growing material. In this context, VHF PE-CVD offers significantly higher deposition rate with a less detrimental effect on the defect density due to the softer plasma and also due to the more efficient dissociation of silane, when compared with "standard" RF PE-CVD [13, 16]. Higher plasma powers also increase the formation of powder in the plasma; these consist of tiny particles which can then be incorporated into the a-Si:H layers and may lead to performance deterioration of the devices (shunts, high leakage currents, higher noise, etc). Here again, higher excitation frequency has a beneficial aspect and leads to a higher threshold for powder formation [13]. In the case of HW, even though no plasma is used and very high deposition rate have been demonstrated, no state-of-the-art devices have so far been fabricated at deposition rate higher than 10 Å/s. Residence time has also been shown to have a clear effect on the defect density. A reduction of the residence time is helpful for achieving high deposition rate, combined with low defect densities [17]. Too low deposition temperatures favor the growth of porous, defect-rich materials, while a too high temperature leads to the evolution of hydrogen out of the film, and, thus, partially suppresses the passivation of the dangling bonds.

Uniformity is an issue for all thin-film deposition techniques. This problem has been solved in a satisfactory manner for the large-area deposition of solar cells; one of the methods for solving it has been to use PE-CVD reactors, or other deposition systems, with moving substrates. For the fabrication of arrays of photodiodes with sizes in the micrometer range and with minimal fixed pattern noise, one only needs "local" deposition uniformity over a very small range; obtaining such conditions has never been reported as a serious problem for the deposition of the a-Si:H layer itself.

Mechanical stress management

For MEMS or for TFA applications requiring thick a-Si:H layers and devices, mechanical stress is a major issue. MEMS require a careful control of the mechanical properties of the material, and, in most cases, layers without any stress or with a slight tensile stress. However, standard state-of-the-art a-Si:H deposited by PE-CVD exhibits a fair amount of compressive stress. The latter depends on the deposition conditions such as plasma excitation frequency, plasma power, temperature, pressure, gas mixture and substrate (due to a possible difference in the thermal expansion coefficients between the substrate and the a-Si:H thin-film layer). As an evolution of the a-Si:H growth is sometimes observed along the growth axis, a stress gradient may appear.

For the control of mechanical stress, one can either change the deposition conditions or treat the sample after deposition. Deposition temperature is, in this context, the key parameter: A lowering of the deposition temperature increases the hydrogen content, reduces the material density and the compressive stress, but also decreases the quality of the layer (the latter effect is manifest for temperatures below 150-180°C). Deposition temperature may also be varied, in order to play around with stress effects caused by the differences in the thermal expansion coefficients between substrates and a-Si:H layer. However, this method, that can lead to a compensation of mechanical stress, only works if the temperature coefficient of the substrate is lower than that of a-Si:H.

Post-deposition treatments consist in a thermal annealing of the sample to drive hydrogen out of the layer as stress is a function of the hydrogen content in the film. The latter can be reduced by an annealing operation at a temperature between 350 and 500°C for several hours [18]. However, due to the relatively high process temperature, this treatment is only practical in a restricted field of applications.

Variation of plasma excitation frequency is a powerful tool for stress control, as shown in Fig. 2. VHF PE-CVD allows for the deposition of stress-free a-Si:H layers at low temperatures (around 170°C); these temperatures are, however, still high enough, so as to obtain satisfactory material quality. Note that stress values, as presented in Fig. 2, are also dependent on the other deposition conditions such as plasma power, pressure, gas mixture, etc.

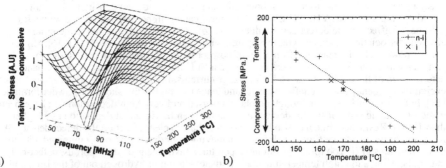

a) b)

Figure 2. Internal mechanical stress in a-Si:H layers deposited by VHF plasma from "pure" silane (no hydrogen dilution): (a) as a function of plasma excitation frequency [19] and (b) as a function of temperature for both intrinsic a-Si:H layers and n-doped-intrinsic a-Si:H bi-layers deposited at 70 MHz (new results).

VERTICALLY INTEGRATED a-Si:H PHOTODIODES

<u>Photodiodes design</u>

In most practical cases, a-Si:H photodiodes use the n-i-p configuration (meaning that the n-layer is deposited first on the substrate – in our case, on the CMOS circuit) with light entering through the p-layer. In a-Si:H solar cells, light has to enter through the p-layer in order to have holes generated as close as possible to the p-layer (this is necessary because the holes exhibit significantly lower collection lengths than electrons). For sensor applications this requirement is not necessary as collection may be enhanced with the applied electric field. Nevertheless, as optimized p-doped layers are usually deposited at slightly lower temperatures than the i-layer, it is still advantageous to use the n-i-p configuration with the p-layer deposited last and with light entering through the top layer, i.e. through the p-layer.

In order to enhance signal to noise ratios, one has here to maximize the external quantum efficiency (QE) while minimizing the dark (leakage) current. A typical QE curve of a-Si:H p-i-n junctions attains values above 80% at the maximum of the curve around 550 nm. As seen in Fig. 3, the value drops below 500 nm; this is due to absorption in the p-layer, where basically no photogeneration takes place: a-Si:H doped layers are very defective and can be considered as dead layers. Therefore, p-layer thickness and doping should be optimized in a way as to minimize optical loss while keeping a blocking contact for low current leakage. QE in the 400 to 500 nm region may be further increased by using p-type a-SiC:H.

A further reduction in QE may be given by the transparent conductive oxide (TCO) used as top contact; the thickness and reflectivity of the TCO layer has to be adapted to the spectral range of interest.

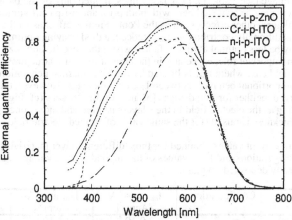

Figure 3. Typical external quantum efficiency (QE) curves of an a-Si:H n-i-p diode compared to a p-i-n diode with a relative thick n-layer and 2 diodes using the metal-i-p configuration. Diodes with ITO top contacts are optimized for maximum QE at 550 nm. Texture ZnO adds an antireflective effect which widen the spectral range. All diodes are 1 µm thick.

Above 600 nm, QE rapidly decreases to almost 0 at 750 nm due to the fact that one approaches here the band gap of a-Si:H (1.75 eV). This part of the QE depends strongly on the thickness of the diode. The sensitivity in the infrared region may be enhanced by using a-SiGe:H alloys. Alternatively, hydrogenated microcrystalline silicon may be used. This material, which has a band gap of 1.1 eV, can be deposited using the same techniques as used for a-Si:H, but under higher dilution of silane with hydrogen. However, the use of lower band gap materials generally leads to higher thermal generation of carriers and, consequently, to larger values of dark current.

Dark leakage current I_{dark} is mainly controlled by carrier thermal generation and carrier injection from the doped layers. As the illumination or as the applied field on the diode is modified, occupation of the localized states in the band gap is affected. (a-Si:H does not strictly have a band gap since it exhibits a continuous distribution of states in the gap). The discharging of localized states leads to time-dependent current decays and tens of minutes may be necessary to reach steady-state conditions [20].If doping of the doped layers (and especially the p-layer) is not sufficient, the discharging of the defects will allow the electric field to extend further into the p-layer and eventually leads to current injections [20]. Because of this time–dependent behavior, dark I(V) characteristics have to be measured with great care, allowing enough time for current stabilization.

The electric field is not distributed uniformly through the intrinsic layer, but is concentrated at the interfaces between the doped layers and the i-layer. It is further not distributed symmetrically but has its highest strength at the p/i interface; this behavior is more prominent for thicker diodes [21]. For this reason, besides the contribution of carrier thermal generation (including the Poole-Frenkel effect, i.e. including field-assisted thermal generation) to I_{dark}, most of the additional effects are due to field injection at the p/i interface. As the n-layer plays a minor role, one can even build reasonable diodes without any n-layer at all; these diodes have an (almost) ohmic contact between the metallic back contact and the i-layer (metal-i-p structure).

Taking into account only the basic thermal generation of carriers (in absence of a strong electric field), one can expect I_{dark} values of the order of 1 pA/cm^2 [22]and this is also what is experimentally observed on optimized n-i-p diodes deposited on glass, at low values of the applied reverse voltage (see Table 1). Low I_{dark} values, limited only by the basic thermal emission of carriers may be achieved here with both p-i-n and n-i-p configurations. However, at higher reverse voltages, other effects such as the Poole-Frenkel effect as well as further current injection effects additionally contribute to I_{dark}. Indeed, a diode having optimal behavior at low electric fields may be less optimal at higher fields, where the specific optimization of the p/i interface plays an important role, especially in the case of metal-i-p structures.

For the low-field case where I_{dark} is limited by the basic thermal generation of carriers, one would expect a proportional behavior between I_{dark} and thickness. However, this relation is usually not observed, neither for thin devices [5], nor for thicker ones (cf. Fig. 4). The latter case is not too surprising as the electric field in the i-layers is more and more concentrated at the p/i interface as the thickness is raised (for the same value of applied electric field).

Table 1. Best dark current values obtained for three different i-layer materials (a-Si:H), for various diode configurations and for 3 values of the applied reverse voltage. All diodes are 1 µm thick and are directly deposited on glass.

i-layer material	Configuration	I_{dark} at -1 V [Acm^{-2}]	I_{dark} at -3 V [Acm^{-2}]	I_{dark} at -5 V [Acm^{-2}]
Low rate 1 (3 Å/s)	n-i-p	$2.0 \cdot 10^{-12}$	$3.6 \cdot 10^{-12}$	$>10^{-10}$
	p-i-n	$1.0 \cdot 10^{-12}$	$3.0 \cdot 10^{-12}$	$6.5 \cdot 10^{-12}$
	metal-i-p	$5.5 \cdot 10^{-11}$	$2.4 \cdot 10^{-10}$	$>10^{-9}$
Low rate 2 (3.3 Å/s)	p-i-n	$9.1 \cdot 10^{-12}$	$2.2 \cdot 10^{-11}$	$2.7 \cdot 10^{-11}$
	metal-i-p	$4.4 \cdot 10^{-12}$	$4.0 \cdot 10^{-11}$	$6.1 \cdot 10^{-11}$
High rate (15.6 Å/s)	n-i-p	$6.1 \cdot 10^{-12}$	$1.6 \cdot 10^{-11}$	

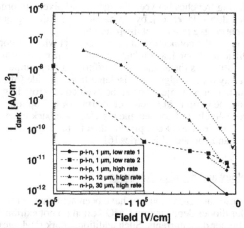

Figure 4. Dark current I_{dark} as a function of reverse bias field and thickness for a-Si:H diodes deposited at low and high rates on glass substrate. Low rate materials are deposited between 3and 3.3 Å/s (using different hydrogen dilution of silane) while high rate material is deposited at 15.6 Å/s [26].

Light-induced degradation (known as Steabler-Wronski effect [23]) has also to be addressed for some applications. This reversible effect is due to the creation of additional metastable defects during light soaking; it has been extensively studied for solar cell applications. For the latter, the effect is usually minimized by working with thin cells in order to maximize the internal field and avoid collection problems. However, reduction of the cell thickness requires the implementation of light-trapping schemes, in order to still obtain sufficient absorption of the incoming light. Such light trapping schemes are often not usable for detectors. Luckily for detector applications, collection should a priori not be a major problem since collection is enhanced by the application of a strong external electric field. However, intensive illumination may still create high enough defect densities, so as to affect collection, to create additional cross-talk and to increase I_{dark} [24].As light-induced degradation is an intrinsic feature of a-Si:H, which cannot be fully avoided in this material, one has to optimize the thickness of the diode and the value of the applied field, and to adjust the design of the detector array in order to minimize the effects provoked by the degradation.

<u>a-Si:H diode integration</u>

Integration of an a-Si:H diode array on a CMOS circuit requires a reliable electrical connection between each diode and the corresponding entry point on the circuit, whilst at the same time guaranteeing an adequate electrical isolation between adjacent pixels, so as to obtain the specified resolution. The a-Si:H diode array is deposited directly on the dielectric layer of the circuit: this fact ensures a high degree of isolation between the diodes and the electronics. Back contacts of the diodes are defined by metallic pads which are either a part of the CMOS chip itself (CMOS metal layer) or are deposited on the chip (vias or pads) and subsequently patterned. A sequence of continuous, uninterrupted a-Si:H layers is then deposited to form the diode structure (see Fig. 1). The advantage of such a continuous layer arrangement is that dead areas can be almost fully eliminated. When using this simple integration scheme, all back contact pads are connected through the (continuous, uninterrupted) back doped layer (usually the n-layer); and this fact may induce significant cross-talk. This effect can be reduced by a patterning of the back doped layer, by introducing trenches between the individual pixels (in order to have a discontinuity in the back doped layer) [4], by a reduction in the doping of the back doped layer or by completely eliminating the latter (metal-i-p structure).

Another crucial point is the connection to the top contact which is common for all pixels. As TCO layers are very difficult to bond, a conductive bridge is needed from the TCO layer to a chip via or pad. This solution has the advantage of offering a monolithic connection but will inject current into the array if the bridge is not isolated from the side (of the array). The detrimental effect of such a current injection can be avoided by providing an additional guard ring [5] or by sacrificing the peripheral pixels of the array. A last option is to get rid out of the common top contact completely and connect the top electrode back to an adjacent ring metal pad through the same a-Si:H diode array. Each pixel is then formed by a device consisting of two diodes facing each other and connected in series [6].

CMOS circuit morphology and diode performance

Even though very low dark current values have been achieved in test structures deposited directly on glass, similar diodes deposited on CMOS often do not exhibit the same performance, but have significantly higher dark currents. Such additional dark (leakage) current becomes thereby one of the most crucial points forTFA technology, when used for light detection. The problem originates in almost all case from edges or tips where electric field concentration occurs [6, 22, 25]. An example of this leakage effect is shown in Fig. 5. The most obvious solution is to use planarized CMOS technology. However, access to planarized technologies is not always possible (for financial or technical reasons) and one has to find other ways of reducing the leakage effects. Minimizing edge effects is in most cases possible by opting for a large opening of the passivation around the pads, or, more drastically, by having no passivation of the chip above the last metal layer [22], by etching the passivation layer in order to have a smoother morphology [6] or by using metal-i-p structures [1, 22]. The possible morphologies are schematically drawn in Fig. 6. In each case, one can have then: either (a) a continuous deposition of a n-i-p a-Si:H diode, or (b) a patterned n-layer and a continuous i-p structure, or (c) a continuous i-p structure, without any n-layer (metal–i-p structure). For non-planarized technologies, the lowest dark currents are obtained for the smoother chip morphologies, combined with the metal-i-p structure [22].

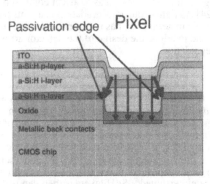

Figure 5. Schematic cross view of an individual pixel of a TFA sensor, with an a-Si:H array deposited on a non-planar CMOS chip, illustrating the resulting peripheral leakage currents.

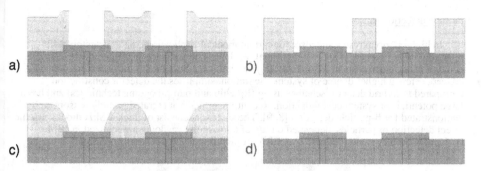

Figure 6. Various types of morphologies for the opening in the passivation around the pixel: (a) standard CMOS pad, (b) large opening in the passivation, (c) partial etching of the passivation (d) unpassivated or planarized CMOS technology.

Visible light sensors

   To illustrate the effect of chip morphology, several visible light imagers were fabricated with a 1 µm thick a-Si:H diode array both in the metal-i-p and in the n-i-p configurations, using standard passivated chips as well as unpassivated ones (cf. Fig. 7) [26]. CMOS chips provided by CSEM consisted in an array of 64x64 pixels, with a pixel lateral size of 33 µm (passivated chip) or 38 µm (unpassivated) and a pitch of 40 µm in Alcatel-Mitag 0.5 µm MPW technology. Each pixel is connected within the CMOS chip to an individual charge integrator.
   Dark current values with the a-Si:H diodes for the various configurations are given in Table 2. A very low value of 21 pA/cm^2 is obtained, at a polarization of -1 V, for the best sensor, i.e. when using a metal-i-p configuration on an unpassivated chip [22]. Radiometric sensitivity values between 20 and 80 V/(µJ/cm^2) have been obtained, for these sensors, at a wavelength of 500 nm.

Table 2. Dark (leakage) current values of individual pixels at a reverse bias of -1 V, obtained for 1 µm thick a-Si:H diodes, in n-i-p or metal-i-p configurations, deposited on passivated or unpassivated CMOS readout chips.

Diode structure and type of chip	Dark (leakage) current [A/cm^2]
n-i-p on passivated chip	2.5×10^{-9}
metal-i-p on passivated chip	3.6×10^{-10}
metal-i-p on unpassivated chip	2.1×10^{-11}

Figure 7. View of a TFA imager with 64x64 pixels, with a lateral pixel size of 33 µm and a pitch of 40 µm, on which a 1 µm thick a-Si:H diode array has been deposited.

<u>Particle detection sensors</u>

In high-energy physics and also, for certain medical applications, pixel detectors with relatively large pixel areas and with minimal dead areas between the pixels are desired. In this context, particle detectors in TFA technology with thick a-Si:H diodes are attractive. The TFA concept offers a higher degree of system integration, simplifies the detector construction (compared to hybrid detector schemes using flip-chip and bump-bonding techniques) and has a large potential for system cost reduction. Recently, vertically integrated particle sensors were demonstrated for β-particle detection [8, 9]. These sensors consist of thick a-Si:H diodes, for the direct detection of particles, deposited on top of a low-noise readout chip realized in IBM 0.25 μm technology and provided by CERN. A picture of such a sensor is presented in Fig. 8.

Figure 8. View of a 8x6 pixel particle detector in TFA technology with a 15 μm thick a-Si:H diode array. Each octagonal pixel has a lateral size of 150 μm with a pitch of 380 μm.

For this application, a first important issue is given by fabrication/deposition time, as here a-Si:H diode arrays with thicknesses ranging between 15 and 30 μm have to be deposited, in

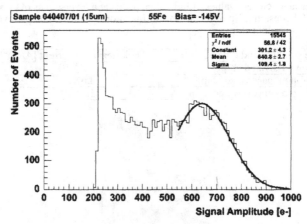

Figure 9. Spectrum of 5.9 keV photons from a ^{55}Fe source, as obtained with the sensor presented in Fig. 9. The 15 μm thick a-Si:H diode array is operated at a reverse bias voltage of 145V. The peak at low energy values is due to noise and was cut off below 200 electrons

order obtain enough electron hole pairs for charge collection . The second issue (related also to diode thickness) is to achieve full depletion of the diode, by applying a high enough reverse bias voltage, and still limit the dark (leakage) current to a reasonable value [8]. VHF PE-CVD is here very well suited for diode fabrication, as with this method, high quality a-Si:H with low internal stress can be deposited at high rates. Diodes with thicknesses of up to 32 μm were deposited at 15.6 Å/s using a plasma excitation frequency of 70 MHz, for the sensor application mentioned above.

Single-particle detection at the minimum ionizing energy (MIP) has not yet been demonstrated with β particles; this is due to the insufficient signal to noise ration so far achieved [9]. Further progress is, in fact, needed here to achieve full depletion in very thick diodes while keeping acceptable leakage current values. On the other hand, detection of X-rays have been successfully demonstrated (cf. Fig. 9) [27], opening up novel perspectives for X-ray imaging.

CONCLUSIONS

Integration of sensors together with their readout electronics is increasingly becoming a challenge, especially with the decreasing feature size in microelectronics. Not only the floor space of the chips is also decreasing, but the coupling of the sensors with the environment becomes problematic. A vertical integration of the sensor on the electronic circuit, using thereby thin-film silicon sensors, is an obvious solution to the problem.

For light or particle detection or imaging, amorphous silicon (a-Si:H) thin-film devices deposited on CMOS readout chips (so-called TFA or TFC technology) offer indeed an attractive alternative to CMOS APS. Their main advantage resides in the high "geometrical" fill-factor (i.e. high ratio of active area to total area), and in the reduction of dead areas. For application to particle detection in high energy physics, the radiation hardness of a-Si:H and the resulting highly simplified integration scheme for the TFA sensor (which contrasts with the complexity of present hybrid technology) are two additional valuable assets. TFA sensors are, thus, very well suited for applications requiring very high sensitivity, full area sensitivity, or for situations involving a complex CMOS readout circuit, which would otherwise limit the area devoted to the sensor.

However, the development of a-Si:H sensors on CMOS readout chips makes it mandatory to address several additional issues that concern thin-film deposition, thin-film device design, exact geometrical configuration for integration, and also the details of the electrical connections between the thin-film diodes and the CMOS chip. It is only by solving the related problems that optimal performances will be obtained. Even though vertical integration basically permits one to separate the circuit design from the sensor optimization, the reduction of feature size may cause, unless special care is taken, a significant increase in sensor dark (leakage) current due to peripheral effects and, thus, have negative consequences on device performance. The present paper has shown a number of potential solutions to avoid these negative consequences. These solutions have been successfully tested on individual experimental chips, achieving thereby a dark current value of 21 pA/cm^2. The next step will be to introduce them in actual products.

Using the same technology, ionizing particle and X-ray detectors using up to 32 μm thick a-Si:H diodes deposited on low noise readout chip have been successfully fabricated and tested. With the high radiation hardness of a-Si:H, this novel type of sensor are very attractive for applications in high energy physics experiments, as well as in non-destructive testing systems.

ACKNOWLEDGEMENTS

The authors acknowledge financial support from CSEM through the CSEM-IMT joint research program and from CERN.

REFERENCES

[1]   S. Benthien, T. Lulé, B. Schneider, M. Wagner, M. Verhoeven, M. Böhm, *IEEE Journal of Solid State Circuits* **35**, 939 (2000).

[2]   J. Sterzel, F. Blecher, M. Hillebrand, B. Schneider, M. Böhm, Mater. Res. Soc. Proc. **715**, A.7.1.1 (2002).

[3]   T. Lulé B. Schneider, M. Böhm, IEEE. *J. of Solid-State Circuit* **34**, 704 (1999).

[4]   B. Schneider, P. Rieve, M. Böhm.,in B. Jähne, H. Haußecker, P. Geißler, Handbook on Computer Vision an Applications, Academic Press, Boston, pp. 237-270, 1999.

[5]   J. Theil, *IEE Proc. Circuits Devices Syst.* **150** 235 (2003).

[6]   T. Neidlinger, C. Harendt, J. Glockner, M.B. Schubert, Mater. Res. Soc. Proc. **558**, 285 (1999).

[7]   A.J. Syllaios T.R. Schimert, R.W. Gooch, W.L. McCardel, B.A. Ritchey, J.H. Tregilgas, Mater. Res. Soc. Proc. **609**, A14.4.1 (2001).

[8]   N. Wyrsch, C. Miazza, S. Dunand, A. Shah, D. Moraes, G. Anelli, M. Despeisse, P. Jarron G. Dissertori, G. Viertel, Mater. Res. Soc. Proc. **808**, 441 (2004).

[9]   N. Wyrsch, S. Dunand, C. Miazza, A. Shah, G. Anelli, M. Despeisse, A. Garrigos, P. Jarron, J. Kaplon, D. Moraes, S.C. Commichau, G. Dissertori, G.M. Viertel, *Physica Status Solidi (c)* **1**, 1284 (2004).

[10]   T, Lulé et al., *IEEE Trans. on Electron Devices* **47**, 2110 (2000).

[11]   N. Wyrsch, C. Miazza, S. Dunand, C. Ballif, A. Shah, M. Despeisse, D. Moraes, P. Jarron' to be published Mater. Res. Soc. Proc. **862**.

[12]   B. Rech, S. Wieder, F. Siebke, C. Beneking, H. Wagner, MRS Symp. Proc. 420 (1996) 33.

[13]   A. Shah et al., Mater. Res. Soc. Proc. **258**, 15 (1992).

[14]   L. Paquin, D. Masson, M.R. Wertheimer, M. Moisan, *Canadian Journal of Physics* **63**, 831 (1985).

[15]   A. Mahan, Proc. of the 3rd World Conf. on Photovoltaic Energy Conversion, 1556 (2003).

[16]   R. Platz, S. Wagner, C. Hof, A. Shah, S. Wieder, B. Rech, *J. Appl. Phys.* **84**, 3953,(1998).

[17]   A. Matsuda, *J. Vac. Sci. Technol.* **A16**, 365 (1998).

[18]   C. Yeh, J.B. Boyce, J. Ho, R. Lau, Mater. Res. Soc. Proc. **609**, A21.3.1 (2000).

[19]   P. Chabloz, H. Keppner, D. Fischer, D. Link, A. Shah, *J. Non-Cryst. Sol.* **198-200**, 1159 (1996).

[20]   R. Street, *Phil. Mag. B* **63**, 1343 (1991).

[21]   J.B Chévrier, B. Equer, *J. Appl. Phys.* **76**, 7415 (1994).

[22]   C. Miazza, N. Wyrsch, G. Choong, S. Dunand, A. Shah, C. Ballif, R. Kaufmann, F. Lustenberger, N. Blanc, M. Despeisse, P. Jarron, this volume.

[23]   D.L. Staebler, C.R. Wronski, *Appl. Phys. Lett.* **31**, 292 (1977).

[24]   J.A. Theil, Mater. Res. Soc. Proc. **808**, 435 (2004).

[25]   C. Miazza, C. Miazza, S. Dunand, N. Wyrsch, A. Shah, N. Blanc, R. Kaufmann, L. Cavalier, Mater. Res. Soc. Proc. **808**, 513 (2004).

[26]   N. Wyrsch C. Miazza, S. Dunand, A. Shah, N. Blanc, R. Kaufmann, P. Jarron, M. Despeisse, D. Moraes, G. Anelli, Mater. Res. Soc. Proc. **762**, 205 (2003).

[27]   D. Moraes et al., Proc. of the 10th Workshop on Electronics for LHC Experiments and Future Experiments, Boston, 2004.

Mater. Res. Soc. Symp. Proc. Vol. 869 © 2005 Materials Research Society D1.2

Influence of design parameters on dark current of vertically integrated a-Si:H diodes.

C. Miazza[1], N.Wyrsch[1], G.Choong[1], S. Dunand[1], C. Ballif[1], A. Shah[1], N. Blanc[2], F. Lustenberger[2], R. Kaufmann[2], M. Despeisse[3], P. Jarron[3]
[1] Institut de Microtechnique, Université de Neuchâtel, Breguet 2, 2000 Neuchâtel, Switzerland,
[2] CSEM SA, Badenerstrasse 569, P.O. Box, 8048 Zurich, Switzerland,
[3] CERN, CERN Meyrin, 1211 Geneva 23, Switzerland.

Abstract

Image and particle sensors based on thin film on CMOS (TFC) technology, where a-Si:H detectors are vertically integrated on top of a CMOS chip, basically provide high sensitivity and low dark current densities (J_{dark}). However, as shown in previous work and as confirmed by the actual measurements, J_{dark} values depend on the topology of the chip and on the detector structure used.

The present paper describes a systematic study carried out, both with test structures on glass and also with a dedicated CMOS test chip designed by CERN. The increase in J_{dark} is shown to be related to border effects, and especially on the detailed structure of the pixel periphery. In all cases, lower J_{dark} are obtained when one uses metal-i-p instead of n-i-p configuration detectors. Transferring these results to the standard TFC sensors used by them, the authors have obtained values of J_{dark} as low as 20 pA/cm^2 at -1 V reverse bias.

Introduction

The vertical integration of a thin-film a-Si:H detector on top of a dedicated CMOS integrated circuit is an attractive solution to enhance the performances of the resulting monolithic sensors. This approach is useful in the field of visible light imaging [1, 2], particle detection and X- or γ-ray detection [3, 4, 5].

For monolithic image sensors a high sensitivity (> 50 V/ (µJ/cm^2)) together with potentially low dark current densities (J_{dark}) can be achieved (<10^{-10}A/cm^2). In fact, TFC (Thin Film on CMOS) technology, where the pixel readout-electronics does not share the die area with the photodiode array allows one to reach very high geometrical fill factors (FF = A_{eff}/A_{pix} > 90 %). This fact combined with the high quantum efficiency (QE(λ)) of a-Si:H in the visible spectral range improves the final sensitivity [6]:

$$S_w(\lambda) = \frac{FF \cdot A_{pix}}{C_{int}} \cdot \frac{\lambda q}{hc_0} \cdot QE(\lambda) \quad [V/(\mu J/cm^2)] \tag{1}$$

where A_{pix} is the whole pixel area, C_{int} the integration capacitance, λ the wavelength, q the elementary charge, hc$_0$ is the product of Plank constant and the light velocity in vacuum.

In order to reach an enhanced dynamic range and the possibility of very low light level detection a very low J_{dark} values is a key issue [7]. This is also the case in the field of particle detection and X- or γ-ray sensors. Here, depending on the application, one needs to be able to detect as low as 60 to 80 electron/hole charge pairs generated per micron of material thickness. Therefore, it is crucial to understand the origin of J_{dark} and the mechanisms influencing it in a TFC configuration, so as to be able to fabricate vertically integrated monolithic sensors for these applications.

In the case of a-Si:H n-i-p photodiodes the lower limit for J_{dark} is given by the thermally generated charge in the intrinsic layer. The density of thermally generated current depends on the defect density and the bandgap of the semiconductor material [8, 9]:

$$J_{th} = q.d_i.N_{db}.\frac{1}{\tau_{gen}}.\exp[-(E_G)/2kT] \quad [A/cm^2] \tag{2}$$

where kT is the temperature and Boltzmann constant product, d_i the intrinsic layer thickness, N_{db} the dandling bond density, E_G the "equivalent band gap" of a-Si:H and finally τ_{gen} the time constant governing thermal generation in the intrinsic layer.

For a 1 μm thick diode with "good quality" a-Si:H ($N_{db} = 10^{15}$ cm^{-3}) and considering its high "equivalent band gap" ($E_G = 1.8$ eV) one is able to estimate the thermally generated current density (J_{th}). One obtains J_{th} values in the range of 1 to 10 pA/cm^2. Such low values were confirmed by experimental measurements of J_{dark} on millimeter size test structures deposited on glass [10, 11]. However, difficulties in reaching similar low J_{dark} on monolithic TFC sensors were encountered [11]. The reason for this is that additional "extrinsic" sources of leakage current are present. First of all, a size effect was observed for J_{dark} [10]. This behaviour could be related to peripheral pixel border effects.

Similar peripheral currents were also observed in another work in which the authors suggested a dominating injection phenomenon at the p-i interface [12]. This was evidenced for sufficiently small sensors and was strongly dependent on voltage. In our work the peripheral current appears to be different. In fact, the border effects are found to be related to the particular pixel architecture and the surface topology of the underlying CMOS chip.

In order to eliminate or at least minimise these border effects, responsible for the observed increase of J_{dark}, we have to investigate here the influence of the design parameters. With this goal in mind, special test structures were developed and fabricated, both on glass by photolithography and on a dedicated "CMOS test chip" designed by CERN. Thanks to these structures we could gain in testing flexibility compared to the standard TFC sensors.

The results of this study are presented and a crucial factor influencing J_{dark} is evidenced. This leads to a better understanding of the mechanisms governing J_{dark} and give us a design rule together with an alternative solution, in order to reduce J_{dark}. Finally, this knowledge has been transferred to the standard TFC sensors and the measurements obtained thereby confirm the reduction of J_{dark} values also in this case.

Figure 1: Schematic side view (left) and top view (right) of the pixel configuration. In the overlap 20 μm case the passivation covers a part of the metallic layer and is only opened inside the pad. The middle case corresponds to an opening with the same size as the metal pad. The bottom case shows an opening of the passivation layer 20 μm larger than the metallic contact.

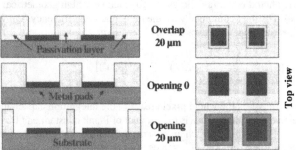

Experimental details

All measurements and results reported in this paper have been obtained on a-Si:H devices deposited in a Very High Frequency Plasma Enhanced Chemical Vapor Deposition (VHF PE-CVD) reactor, at a frequency of 70 MHz and temperature of 200 °C. For the deposition of the intrinsic a-Si:H layers, a concentration of silane (SiH$_4$) of 43 % was used (C = [SiH$_4$]/[H$_2$ + SiH$_4$]). The deposition rate of the intrinsic layer is around 3 Å/s. In the case of the doped layers, diborane (B$_2$H$_6$/H$_2$) for the p layer and phosphine (P$_3$H$_2$/H$_2$) for the n layer were added.

In order to study the effect of pixel geometry and the surface topology, specially designed test structures, so called "chip-like structures", with small-size pixels (40 to 1000 μm side lengths) were fabricated by photolithography. These pixels consist of a-Si:H diodes deposited on quartz wafers "mimicking" the pixel configuration of a CMOS chip. They have back metallic contacts in wells surrounded by a SiO$_2$ isolating layer (fig. 1), with a common top electrode. A design feature that is varied is the oxide opening area. Three different pixel configurations are realized. In a first configuration the passivation covers a part of the metallic layer and is opened only inside the pad (overlap 20 μm, Fig. 1). The second case corresponds to an opening with the same size as the metal pad (opening 0, Fig. 1). And finally an opening of the passivation layer 20 μm larger than the metallic contact (opening 20 μm, Fig. 1) was investigated.

A dedicated "CMOS test chip" has been designed by CERN in order to investigate the influence on J$_{dark}$ of some of the design parameters described above in the case of a vertical integration over it. The pixels implemented in this chip have an octagonal shape and a surface of 0.02 mm^2. The passivation layer of this IBM 0.25 μm process is made of about 6- 7 μm thick polyimide layer over 1 μm thick SiO$_2$ and Si$_3$N$_4$ layers. The pixels present different openings. The first case is an overlapped configuration. The second case exhibits a polyimide layer opened outside the metallic area (15 μm between the polyimide and the metal pad). A third case is a larger opening (40 μm). The standard TFC sensors were also used (see below).

Results and discussion

The first experiment has been to measure the dark current density J$_{dark}$ on the "chip-like structures" for pixels with different opening configurations. As shown in Fig. 2, in the case of a 1 μm thick n-i-p diode with 0.6 μm of passivation, a clear effect is evidenced. The values of J$_{dark}$ measured for pixels in the overlapping configuration are higher than those exhibited by the pixels in the opening case. The explanation of this behaviour is that due to the presence of an oxide step a border effect is present. This border effect causes a peripheral leakage current which is added to the thermally generated current density (equation 2). In the overlapped case, the peripheral leakage current is collected through the <n> conducting layer of the diode. That is why such a high J$_{dark}$ is observed in this case. For an opening of 20 μm, the step is pushed away from the metal contact; therefore the collection of the undesirable leakage current is reduced due to this distance.

In the case of an opening of the same size as the metal pad (opening 0, fig.1) the results were spread over a large range. In fact some values of J$_{dark}$ were close to the one measured in the configuration with 20 μm opening and other values were as large as the ones in the case the overlap. We suspect that misalignment during the photolithographic steps can explain this phenomenon. Therefore these measurements are not taken into account and for further investigation a new design with different opening and overlap lengths will be realized.

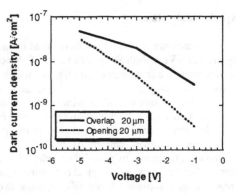

Figure 2: Dark current density in function of the applied voltage measured on a "chip-like structure" with 0.6 μm thick SiO_2, for an n-i-p diode and in the cases of a 20 μm overlap and opening. With an opening configuration pixel J_{dark} is clearly reduced.

Measurements obtained on the test chip designed by CERN and shown in Fig. 3 confirm that there is an influence of the opening length on J_{dark}. Indeed, for n-i-p diodes vertically integrated on the "CMOS test chip" the behaviour is similar as the one described above for "chip-like structures". In fact the overlap case clearly exhibits the highest J_{dark}. However the additional information is that a larger opening is positive; the J_{dark} value measured for a 40 μm opening is lower than that for a 15 μm opening.

After these experiments it was decided to transfer the knowledge to the standard TFC sensor chip so as to further confirm the results. Thereby, several monolithic TFC sensors have been fabricated. The CMOS chips used for this purpose have 32 x 64 pixels of 40 μm side length. The metal contact side length is 38.4 μm. Here tow types of chips were available.

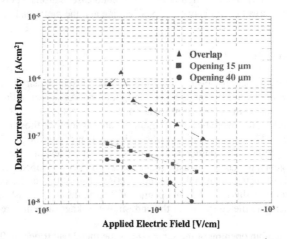

Figure 3: Dark current density in function of the applied electric field (-10^4 V/cm is equivalent to a 1 V reverse bias voltage on 1 μm thick diode). These results have been measured on the CERN test chip with a 1 μm thick n-i-p diode. The pixel in the overlapped configuration (Fig. 1) exhibits the highest dark current density. The case of an opening of 40 μm size shows the lowest dark current density. The border effects, responsible for the increase of J_{dark}, seem to be reduced by a larger opening in the polyimide around the metal pad.

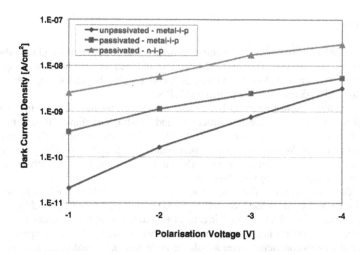

Figure 4: Dark current density in function of the applied voltage measured on TFC monolithic sensors. The metal-i-p diode over the unpassivated chip shows the lower J_{dark} values. The highest values of J_{dark} are measured on a passivated chip with n-i-p diode. Replacing the n-i-p with a metal-i-p on the passivated chip clearly improves J_{dark}. All diodes are 1 µm thick.

A first CMOS chip with 1 µm thick passivation is available in which the opening side length is fixed at 33.4 µm. This means that in this case the pixels are in the overlapping configuration. A second CMOS chip is, on the contrary, unpassivated. This case corresponds to a "global opening".

The best combination, as shown in Fig. 4, is the 1 µm thick metal-i-p configuration diode vertically integrated over an unpassivated CMOS chip (20 pA/cm^2 at -1V). The worst case corresponds to 1 µm thick n-i-p diode deposited on top of a passivated chip. In this case the border effects are dominant. By switching to a metal-i-p diode configuration, as shown by the middle curve (Fig. 4), the border effects can be significantly reduced improving, thus, the values of J_{dark}.

We can conclude that with the metal-i-p diode configuration the absence of bottom <n> conducting layer avoids the collection of the peripheral additional leakage current due to the border effects. Thus, the resulting value of J_{dark} will be much lower and closer to the thermal generation limit J_{th} calculated from equation 2 and verified experimentally on test structures deposited on glass [10, 11].

On the other hand the use of larger openings in the passivation has a similar effect as the use of a metal-i-p configuration for the detector diode. In fact, if the opening is large enough the peripheral leakage current will not be collected thanks to the spacing between the metal pad and the oxide step where border effects take place. This will bring an improvement on J_{dark}; however it will also cause a reduction in sensitivity (see equation 1). Due to the increasing distance between the pixels; the larger the openings will be, the lower the FF will be, reducing, thus, the final sensitivity of the TFC monolithic sensor.

Conclusion

In order to understand the relation between diode configuration and the substrate topology on one hand and the observed behaviour of the J_{dark} on the other, some special test structures have been developed and fabricated, both on glass by photolithography and on a dedicated "CMOS test chip" designed by CERN.

The most interesting result is the clear influence of the opening area in the passivation layer observed on both, "chip-like structure" and "CMOS test chips". In fact, the lowest J_{dark} values were measured in the case of large openings. The overlapping configuration exhibited the highest J_{dark} values. We suppose that the peripheral leakage current resulting from the border effect due to the passivation step is not collected if the opening is sufficiently large. A pixel design with a large opening will therefore help reducing the values of J_{dark}.

This study confirms the conclusion of a previous work [11] in which we suggested to planarize the surface or at least remove the passivation layer of the CMOS chips in order to reach low J_{dark} values. The advantage in using metal-i-p diode configuration as detector in order to get rid off the border effects is also confirmed by the present study. With this type of a diode the absence of an <n> conducting layer avoids the collection of the additional leakage current resulting from the border effects. Similarly large openings, to push the passivation step far enough from the metal pad, can avoid this undesirable collection resulting, thus, also in lower J_{dark} values.

Therefore, we can conclude that a planar CMOS technology is a key issue to obtain a low value of J_{dark} for vertically integrated TFC monolithic photon and particle sensors. In this solution passivation and metal layers come up to exactly the same level and create a smooth planar surface. If one has no access to a planar technology, the opening consideration can be useful to reduce J_{dark} values. However, a large opening will result in a lower FF because of the larger spacing between the pixels and the resulting lower detector area. The sensitivity will, thus, suffer from this choice. The use of a metal-i-p configuration together with unpassivated CMOS chips has shown to be a powerful solution to minimize J_{dark} (20 pA/cm^2 at -1V). So, if a planar technology is not available the latter is the best solution in terms of low values of J_{dark} and high sensitivity.

References

[1]   T. Lulé, B. Schneider, M. Böhm, IEEE J. of Solid-State Circuits 34, 1999, p.704.
[2]   F. Blecher, S. Coors, A. Eckhardt, F. Mütze, B. Schneider, K. Seibel, J. Sterzel, M. Böhm, Int. Conf. on Mechatronics & Machine Vision in Practice, Nanjing, China (2000).
[3]   N. Wyrsch et al., MRS, San Francisco, Vol. 808, 2004, p.441.
[4]   D.Moraes, et al., Journal of Non-Crystalline Solids, 338-340, 2004, p.729.
[5]   M. Despeisse et al., Nuclear Instr. and Methods in Physics Research, A 518, 2004, p.357
[6]   T. Lule et al., IEEE Transactions on Electron Devices, Vol. 47, No 11, 2000, p.2110.
[7]   B. Schneider et al., Handbook on Comp. Vis. and Appl., Ac. Press, Boston, 1999, p. 237
[8]   F. Meillaud, et al., to be published in Solar Energy and Materials.
[9]   R.A. Street, Appl.Phys.Lett.57 (13), pp.1334-1336 (1990).
[10]   N. Wyrsch, et al., MRS, San Francisco, Vol. 762, 2003, p. 205.
[11]   C. Miazza, et al., MRS, San Francisco, Vol. 808, 2004, p. 513
[12]   E.A. Schiff, R.A. Street, Journal of Non-Crystalline Solids, 198-200, 1996, p.1155

Mater. Res. Soc. Symp. Proc. Vol. 869 © 2005 Materials Research Society

REDUCTION OF RESIDUAL TRANSIENT PHOTOCURRENTS IN A SI:H ELEVATED PHOTODIODE ARRAY BASED CMOS IMAGE SENSORS

Jeremy A. Theil[*]

Agilent Technologies, Santa Clara, CA, 95051, U.S.A.
*current contact information: Lumileds Lighting, LLC, MS 91UJ
370 W. Trimble Rd., San Jose, CA 95131, U.S.A., e-mail: jeremy.theil@lumileds.com

ABSTRACT

While a-Si:H based elevated photodiode arrays hold the promise of superior performance and lower cost CMOS-based image sensors relative to those based upon crystalline silicon photodiodes, the one area where a-Si:H based sensor performance has not been as good is in image lag. This problem is only exacerbated by Staebler Wronski Effect induced junction degradation. Image lag is caused by residual charge from photocurrents trapped within the junction once the light source is removed and can be measured for several seconds, even under continuous applied reverse bias. It is seen both in constant and variable bias pixel architectures. However, by carefully controlling a-Si:H junction bias conditions, it is possible to significantly reduce these transient photocurrents. This article will describe how the photocurrent decay time exponent can be reduce by almost an order of magnitude. Finally the physical causes behind image lag in a-Si:H based photodiode arrays will be discussed.

1 INTRODUCTION

Over the last ten years, there has been increasing interest in the use of a-Si:H in photodiode arrays that are monolithically integrated onto integrated circuits [1-3]. Such integration allows a combination of 1) reduced imaging pixel area, 2) reduced sensor cost, 3) lower photodiode leakage, and 4) improved pixel sensitivity. As pixel-level complexity (hence area) grows, the advantages become more apparent. The one area where a-Si:H diode-based pixel performance tends to be deficient with respect to all-crystalline silicon pixels is in image lag. Image lag is typified as a persistent afterimage artifact in the array as a result of charge generated by an earlier measurement (hence the transient photocurrent). There are many systemic causes of image lag in all image sensor technologies, but the high trap-state density of a-Si:H provides an additional mechanism through prolonged carrier emission. In addition, since most a-Si:H diode array architectures use a continuous i-layer to maximize the light gathering area of the array, the junction construction itself may contribute. Therefore we created a novel parametric test structure to try to elucidate various potential mechanisms in a realistic array. The motivation behind this work is to start to identify the causes of this transient photocurrent and how see if there are ways to mitigate it.

2 EXPERIMENTAL

Details of diode fabrication have been presented elsewhere, the resulting structure is pictured in Figure 1a [1,2,4]. All test structures are bounded by a ring diode that is held at the same bias as the measurement structure itself to eliminate injected edge currents. The test structure used in this experiment is a two-channel 2-D interpenetrating diode array consisting of junction area about 8.84×10^5 μm^2 with 940 μm. Such an array mimics layouts for high-density a-Si:H arrays, along with the ability to study the current flow between individual pixels and junctions by aggregating them into larger structures. The n-layer electrode is patterned in a series of 4 μm

square pixels with a 5 µm pitch, in a 188 x 188 array, (see Figure 1b). The two channels each contain 17672 pixels and allow for independent biasing to study interpixel effects. The i-layer thickness was 5500Å, the 200Å thick p-layer boron atomic concentration was 7.0 x 10^{19} cm^{-3}, and the 500Å thick n-layer phosphorus concentration was 2.0 x 10^{20} cm^{-3}, as measured by SIMS. The a-Si:H layers are formed by very high rate PECVD deposition methods (> 30 Å/s), with the resultant films having an intrinsic defect density of < 4 x 10^{15} cm^{-3} [5].

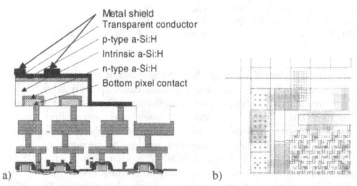

a)   b)

Figure 1: a) Schematic diagram of elevated a-Si:H photodiodes. The metal shield lies over the region between adjacent pixels. It is absent in the control arrays. The transparent conductor overlies the entire structure as well as the p-type contact (P). b) Layout of a corner of the pixel array. The pixels are connected into two interpenetrating checkerboard diode arrays (CTR1, and CTR2) surrounded by an independently biased guard-ring (R1).

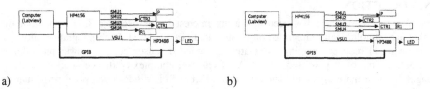

a)   b)

Figure 2: Equipment configuration for the transient photocurrent experiment, a) interpixel transient photocurrent setup, b) junction bias setup.

Measurements were made by pulsing light from a red LED onto the photodiode and monitoring the current decay as a function of time, for various bias configurations using an Agilent 4156B. In one set of experiments, referred to as junction biasing, the bias between the pixels is held to zero, while the overall bias between the common p-type junction and the pixel is varied. For the other set of experiments, referred to as interpixel biasing, the diode is held under reverse bias with respect to the common p-type junction, with various biases applied between the adjacent pixels. A light pulse is generated with a red (Agilent p/n HLMP-EP15, λ_{pk}~630nm) LED connected to an external dc power supply through HP 3488 switch. In these experiments, the Agilent 4156B was put into a sampling mode in which data was collected every 0.1s, and the current was monitored from the point at which the diode bias was switched from 0V to the set bias. All measurements were made in complete darkness and a sample temperature of 21°C.

The transient photocurrent measurements are conceptually quite simple. Once the appropriate biases are applied to the diode, data collection begins, and the light is pulsed. The fall time for the photocurrent is measured as the time from when the LED light is turned off until the value of the photocurrent drops below 1×10^{-11} A. Light pulses of durations from 3 seconds to 30 seconds showed a constant decay behavior, therefore light pulse of 3 seconds were used for the experiments. The LED forward drive bias was set to 2.0V.

3   RESULTS

Figure 3 shows the steady state leakage current through the junction at various electric fields. For the experiments shown here the highest reverse junction bias is 1V, for 5500Å junctions, the 17 pA/cm², or 150fA for a topographically planar junction. Given the field enhancement caused by topography and terminated pixels is about 20x, then the maximum injected current is ~3.0 pA [6], and the edge array injected current is ~1.5 pA, so that the current measured at ICT1R1 in the following figures is predominantly junction leakage and photocurrent.

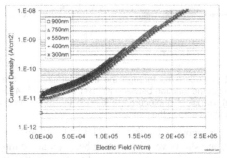

Figure 3: Current density as a function of electric field for an a-Si:H diode series with the same layout but different i-layer thickness.

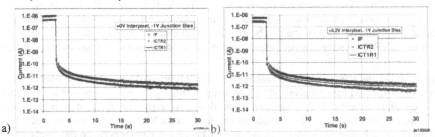

a)   b)

Figure 4: a) Transient photocurrents through the three terminals of the photodiode array, at −1.0V reverse bias across the junction. b) Transient photocurrents through the three terminals of the photodiode array, at −1.0V reverse bias across the junction between ICT1R1 and P, and +200mV bias of ICTR2 w.r.t. ICT1R1.

Figure 4a shows the photocurrent decay as a function of time when the diodes are held under a constant reverse bias. After an initial rapid decline one the light source is removed, the transient response shows the classical logarithmic decay indicative of deep-trap emission, as shown by Wieczorek and Fuhs [7]. In this experiment, the current is evenly balanced between the common P electrode and the two n-type junction channels ICTR2 and ICT1R1. Given that ICT1R1 is also connected to the guard ring, which has been shown to have very high leakage currents through the array edge [1,8], it shows the transient currents are much larger than the injected currents for this timescale and thus may be neglected for the purpose of these experiments. From a practical point, Figure 4 shows the behavior a photodiode array would see under constant voltage circuit operation.

Figure 4b shows the a similar experiment in which CTR2 has a +200mV bias relative to CTR1/R1, and 800mV reverse bias relative to P. Looking very closely between Figure 4a and figure 4b, one can discern that there is a very slight decrease in the current flowing through P, a marked decrease in the current flowing through ICTR2 and an increase in ICTR1/R1.

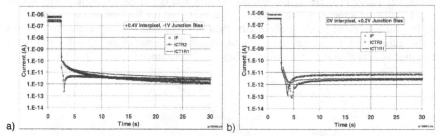

a)   b)

Figure 5: a)Transient photocurrents through the three terminals of the photodiode array, at -1.0V reverse bias across the junction between ICT1R1 and P, and +400mV bias of ICTR2 with respect to ICT1R1. b) Transient photocurrents through the three terminals of the photodiode array, at +0.2V reverse bias across the junction.

Figure 5a shows a more pronounced effect at CTR2 at +400mV relative to CTR1/R1, and 600mV reverse bias to P, and shows very interesting behavior at the three terminals. In this case, one sees that a large steady state current is measurable through ICTR2 after about 1200 msec. The current through CTR1/R1 is now higher than through CTR2, and both have steady state values higher than P. This is interpreted as follows. 1) The initial decline in ICTR2 current is due to a faster reduction in photocurrent, 2) whereas the steady state value for ICTR2 is dominated by and injected current *between* the pixels. This is further evidenced by the magnitude increase seen in the ICTR1R1 current, which appears to be added to the transient photocurrent for this terminal.

Figure 5b shows the effect of holding both CTR2 and CTR1R1 at 0 bias relative to one another, but at a slight forward bias relative to P. Once again in this figure, it is interpreted as an initial drop in the photocurrent, which is quickly replaced with a steady state injected current. In this experiment one sees that each of the photocurrent decay at each terminal is the same and much faster than seen in Figure 4a. In this circumstance, since there is no bias between the pixels, there is no current flow between them, but rather through the junction. As one can see the current through CTR2 and CTR1R1 are equal and both add up to equal the current through P.

Figure 6a shows photocurrent decay plots for three different junction bias conditions. Under reverse bias conditions from −2.0V down to 0V, there is a small but noticeable decrease in the characteristic time for the photocurrent transient, however a forward bias of only 200mV forward bias drastically decreases the decay time. The characteristic decay exponent (τ) of photocurrent found from the slope of a log-log plot of current versus time, for various junction biases is plotted in Figure 6b, and shows a clear transition in τ as a function of bias with the inflection point at 0.0V.

24

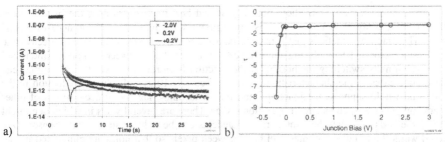

a)

b)

Figure 6: a) Transient photocurrent decay through the pixel as a function of junction bias. Note the large drop in decay time for a small forward bias relative to the small decrease as a function of reverse bias. b) Transient photocurrent decay exponent as a function of junction bias from 3.0 down to -0.2V.

4 DISCUSSION

It is quite likely that the residual photocurrent remaining within the i-layer consists of the trapping of carriers in deep-level states. For the minimum timescale of measurements presented here ($\sim 10^{-1}$s), it can be assumed that the residual charge extracted from the i-layer has been thermalized into deep-level, localized states [9]. The low-defect density (~ 600Å mean defect separation) of the material implies that trapped charge is spatially isolated, since the mean trapping distance is ~ 10Å [10]. Additionally, since thermalization leaves trapped charge in a potential well on the order of several tenths of an eV, recombination between trapped carriers takes place at an extremely low rate [11]. Therefore it is generally believed that trapped carrier recombination occurs through free-carrier capture. Carriers injected into the i-layer initially have high enough energy that they can traverse extended states or shallow localized states attracted towards trapped charge through Coulombic forces.

What is most interesting however, is the phenomenologically different transient decay behavior of the different electrodes during interpixel biasing, which can be thought of as an n-i-n device. When the 400mV interpixel bias is applied across the 1 μm space between pixels, the generated current reduces the photocurrent for the less negative pixel, but increases the photocurrent for the more negative pixel. The interpixel bias diverts a portion of the injected current from more negative pixel, thus reducing the recombination rate. On the other hand, the additional current in the less negative pixel enhances the recombination rate. The current increase in the more negative pixel comes from attraction of a portion of the hole bias, while the decrease in the lower bias pixel current comes from electron transport across the interpixel region. The electron injection is an opportunity for hole annihilation, as there is a probability of EHP recombination.

Field effects and current injection can explain the effect of the junction bias on photocurrent decay. Under forward bias conditions, the magnitude of injected current is quite large and therefore provides a source of carriers for recombination. Counter-intuitively though, while the contribution to reverse bias currents (contact injection, thermal generation, and field–assisted tunneling), increase as a function of reverse bias, it is observed that the photocurrent decay time also increases. One explanation to this paradox is that the increasing electric field tends to decrease the recombination rate by reducing the carrier transit time across the junction. However, as seen in Figure 6b, the effect is weak.

The results of the experiments show both a practical benefit of controlling the bias between different diode array elements, as well as highlighting some interesting phenomena of a-Si:H diodes. The practical benefit is that it may be possible to decrease the characteristic decay time by at least a factor of 10, by placing a few hundred millivolts field between two pixels (interpixel bias). Additionally switching the biasing of the junction from reverse to small forward bias can produce decreases in the characteristic time by as factor of 10 as well. Such a reduction in the decay time would suppress the junction contribution to image lag below decay times of other system components, such as circuit operation issues. There are several ways in which one might imagine methods in which the photocurrent transients can be reduced in an image sensor. One method would be to pulse the interpixel bias between pixels so that holes are eliminated in one portion of the array, the switch the bias sign to eliminate holes in the other portion of the array. Another method would be to apply a brief forward bias pulse to the array to inject enough charge to annihilate the remaining holes, then switch back to reverse bias to extract the remaining electrons.

5 SUMMARY

In this work is has been shown that currents injected into or between elements of an a-Si:H photodiode array modulate photocurrent decay, either by 1) charge transfer of injected currents between pixels, or 2) massive current injection from forward biasing the diode. In both schemes, it is possible to decrease the characteristic decay time by 10x. It appears that the process by which this occurs, is by electron injection and subsequent trapped hole annihilation. The practical result is that it is possible to control the degree of lag currents contributed by the a-Si:H photodiode, so that it is not the limiting factor with respect to lag in CMOS image sensor development.

REFERENCES

[1] J. A. Theil, R. Snyder, D. Hula, K. Lindahl, H. Haddad, and J. Roland, *J. Non-Cryst. Sol.*, **299**, 1234 (2002).

[2] J. A. Theil, M. Cao, G. Kooi, G. W. Ray, W. Greene, J. Lin, A. J. Budrys, U. Yoon, S. Ma, and H. Stork, *MRS Symp. Proc.*, **609**, A14.3.1 (2000).

[3] H. Fischer, J. Schulte, P. Rieve, and M. Böhm, *Mat. Res. Soc. Symp. Proc.*, **336**, 867 (1994).

[4] J. A. Theil, H. Haddad, R. Snyder, M. Zelman, D. Hula, and K. Lindahl, *Proceedings of the SPIE*, **4435**, 206 (2001).

[5] J. Theil, D. Lefforge, G. Kooi, M. Cao, and G. Ray, *J. Non-Cryst. Sol.*, **569** 266 (2000).

[6] J. A. Theil, *Mat. Res. Soc. Symp. Proc.*, **762**, 21.4 (2003).

[7] H. Wieczorek, and W. Fuhs, *Phys. Stat. Sol. (a)*, **109**, 245 (1988).

[8] J. A. Theil, *IEE Proc. Circuits, Devices, and Systems*, **150**(4), accepted for publication (2003).

[9] R. A. Street, *Hydrogenated Amorphous Silicon*, (Cambridge University Press, 1987) pp 288-292.

[10] R. A. Street, *Phys. Rev.*, **B27**, 4924 (1983).

[11] N. Schultz, B. Yan, A. Efros, and P. C. Taylor, *J. Non-Cryst. Sol.*, **266**, 372 (2000).

Thin Film on ASIC (TFA) - A Technology for Advanced Image Sensor Applications

Juergen Sterzel[1], Frank Blecher[2]

[1]Jena-Optronik GmbH, JTE, Pruessingstr. 41, D-07745 Jena, Germany
[2]Lambda Lab, Kohlbettstr. 20, D-57072 Siegen, Germany

ABSTRACT

Thanks to its three-dimensional integration and the use of amorphous as well as crystalline silicon, the TFA technology is suitable for advanced image sensor applications. This paper describes the fundamentals of the properties: sensitivity, dark current, temporal and fixed-pattern noise of these TFA image sensors. It compares the different sensitivity definitions, especially current sensitivity and the charge conversion factor. Further, the dark current sources are pointed out, and their temperature behavior is described. By noise calculations, different pixel input stages are compared with regard to low light level detection.

INTRODUCTION

Developments in the CMOS image sensor technology are widely discussed. Many pixel concepts and new applications have been presented in the last years. A further development of the CMOS image sensor concept is TFA (Thin Film on ASIC) or TFC (Thin Film on CMOS) technology. The thin film is an amorphous silicon layer which is deposited on a crystalline ASIC (figure 1) in a low-cost process. The amorphous layer forms the detector. Several research groups have published application fields for this technology (e.g. [1-3]).

Because of the different physical properties of amorphous and crystalline silicon and the three-dimensional integration, this technology is qualified for advanced image sensor applications. Amorphous silicon has a higher current sensitivity in the visual spectral range, and the technology is much less susceptible to sensitivity decrease after a scaling down of the crystalline ASIC structures [4]. Further, an amorphous detector has a lower dark current than a crystalline detector. And, if different layers are used for the detector and for pixel preprocessing, the readout circuit enables a complex pixel input stage combined with a high sensitivity, because the fill factor remains to 100%. Image sensors for special environments (for example, radiation hard sensors) are also possible.

As it is described, the TFA technology can open new application fields, as its central image sensor parameters can be superior to those of other technologies. This makes TFA eligible for advanced image sensors. However, today's descriptions of central TFA image sensor parameters are inadequate. The amorphous and the crystalline parts are described separately by most authors. This is insufficient, because in a monolithic block the interaction between these two parts is important. Especially for low light level applications (photocurrents in the range of 1 fA) or for high pixel areas (with high integration capacitances), some additional effects have to be considered. Neglecting these, important image sensor parameters are estimated incorrectly.

This paper contains description methods and physical principles important for calculating the parameters of sensitivity, dark current, or temporal noise. It is supplemented by methods to lower detection limits or to increase signal quality. Special consideration is given to dark current temperature dependence. Most of the following descriptions refer to the mainly discussed integrating image sensors.

SENSITIVITY

The detection limit is reached if the ratio between the signal and the signal uncertainties is insufficient. The signal is described by the number of the photo-generated and collected charges, and the sensitivity describes at which time this signal is obtained. Therefore it is assumed, in a first approximation, that advanced image sensors should have a high sensitivity.

There exist different sensitivity descriptions, which differ in the properties included in this description. If the photocurrent is relevant, the current sensitivity S_I is used. This description specifies the transfer of the incident irradiation power to the output current. The voltage sensitivity S_V describes the transfer of the incident light energy to the output voltage [4]. Both descriptions use the ratio between the incident light and an electrical output value. These definitions comprise image sensor parameters like quantum efficiency, fill factor, and the wavelength of the incident light.

A third image sensor sensitivity parameter, which is especially important for integrating image sensors, is the charge conversion sensitivity S_{ch}. This sensitivity, which specifies only the electrical behavior, can be expressed by

$$S_{ch} = \frac{q \cdot v}{C_{int}} \cdot 10^6 \quad \left[\frac{\mu V}{e^-} \right] \tag{1}$$

A high sensitivity S_{ch} is obtained if the capacitance C_{int} is very low and the gain v is very high. Although this definition is very simple, it is sufficient to estimate the effects of the crystalline ASIC on the sensitivity of the whole TFA image sensor.

The TFA technology is a 3-dimensional concept, which results in some parasitic capacitances (figure 1). All these capacitances have to be considered to calculate the integration capacitance C_{int}, of formula (2). The capacitance of the amorphous detector has the symbol C_{det}.

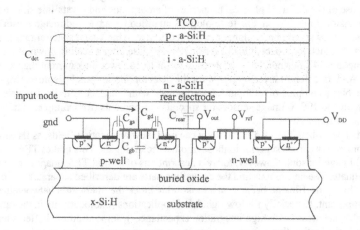

Figure 1. Cross section of a TFA image sensor with SOI substrate and common source pixel input stage for a radiation hard design. Important parasitic capacitances are included.

C_{i0} combines the capacitances, which effect between the input node and the reference point gnd. In figure 1 it consists of the gate-source C_{gs} and the gate bulk capacitance C_{gb}. All capacitances, which effect between the rear electrode of the detector and the output node are combined in C_{oi}. In the example of figure 1 it consists of the capacitances C_{rear} and C_{gd}. Because of the Miller effect, the last term C_{oi} is coupled with the gain v.

$$C_{int} = C_{i0} + C_{det} + C_{oi} \cdot (1 - v), \text{ with } C_{i0} = C_{gs} + C_{gb} \text{ and } C_{oi} = C_{rear} + C_{gd} \tag{2}$$

Formula (2) substituted for C_{int} in formula (1) and a gain v in the range of 1 result in a charge conversion efficiency that is dominated by the detector capacitance added to C_{i0}. A high gain v means that the capacitance C_{oi} dominates. Although these simplifications are used very often, they are erroneous, as shown in figure 2 and figure 3 with state-of-the-art parameters and a detector thickness of 1 μm.

Most image sensors have a source-follower concept as a pixel input stage. The gain of this concept is described with

$$v_{SF} = \frac{g_{m_{driver}}}{g_{ds_{driver}} + g_{ds_{load}} + g_{s_{driver}} + g_{m_{driver}}} \tag{3}$$

and is smaller than 1. g_m describes the gate channel transconductance, g_s the source bulk transconductance and g_{ds} the reciprocal small signal channel resistor. The influence of the different capacitances of a TFA sensor on the charge conversion sensitivity is illustrated in figure 2. If the design is optimized, this sensitivity is dominated by the detector capacitance of the amorphous layer and the capacitance of the reset transistor in the crystalline ASIC. The results of figure 2 are simulated with the parameters of the DMILL 0.8 μm process, lowest possible drawn transistor geometries, and with an amorphous detector capacitance of 10 nF·cm^{-2}. The influence of the reset transistor for lower pixel area is obvious.

On the other hand, if a pixel input stage with a high gain is selected, the coupling capacitance C_{oi} dominates [5]. A simple pixel input stage with high gain is formed by a common source concept and complementary CMOS transistors. The gain of this concept is described with

$$v_{SC} = \frac{g_{m_{driver}}}{g_{ds_{driver}} + g_{ds_{load}}} \tag{4}$$

With this concept gains in the range of up to 200 can be achieved. However, with this gain the capacitances of the reset transistor or the detector should not be neglected, as can be seen in figure 3 (the simulations are done with the same parameters as in figure 2). With a higher pixel area, the detector capacitance increases and the total sensitivity of the pixel decreases. This property will be seen more clearly if the gain of the pixel input stage is small.

Both concepts have shown that the simple assumptions of only the detector capacitance or only the coupling capacitance lead to an overestimation of the sensitivity. Further, the design of the 3-dimensional TFA image sensor has a clear influence on the sensitivity. Low gain and low pixel area have a reduced sensitivity (figure 2), while high gain and high pixel area also result in a decreased sensitivity (figure 3).

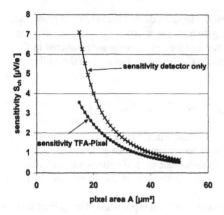 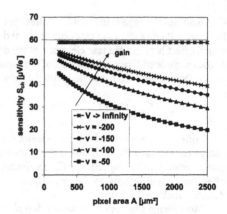

Figure 2. Sensitivity of a TFA pixel input stage with a source follower concept. The upper curve describes the simplification if only the detector capacitance is considered.

Figure 3. Sensitivity of a TFA pixel input stage with the common source concept. The upper curve describes the simplification if only the coupling capacitance is considered.

Besides the absolute value of the sensitivity, the linearity over the whole dynamic range is also important. This linearity is dependent on the conversion factor of the incident photons compared to the photocurrent, termed the quantum efficiency. In a first approximation, this quantum efficiency is dependent on the reflected and absorbed shares in the top layers. Additionally, a constant part of generated charges recombine before they can contribute to the photocurrent. This part affects the linearity, is dependent on the voltage fed to the detector and the density of the dangling bonds. Since this is an offset error, the selected illumination range is also important. Figure 4 shows the photocurrent of a crystalline pn-junction dependent on the reverse voltage. This pn-junction was produced in an AMIS-1.5 μm process. The drawn area has a value of 7320 μm². The used irradiation source was a white LED (T = 8000 K), and irradiation was reduced with grey filters. Low irradiations show the voltage-dependent photocurrent.

In figure 5 the same measurements are plotted vs. illumination. It shows the deviation for lower illumination levels more clearly.

For higher illumination levels, the non-linearity of the integration capacitance C_{int} dependent on the voltage is more important. This capacitance variation of a pn-junction in the crystalline ASIC can be described by

$$C_{pn} = \frac{C_0}{\sqrt{1 + \dfrac{|V_{rev}|}{V_{diff}}}} \tag{5}$$

C_0 denotes the capacitance in the thermodynamic equilibrium, V_{diff} the diffusion voltage, and V_{rev} the voltage across the pn-junction in reverse direction.

30

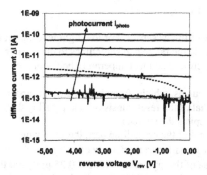

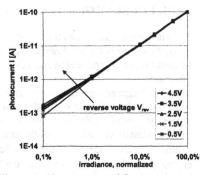

Figure 4. Photocurrent of a pn junction (100%, 50%, 20%, 10%, 1%, and 0.1%), formed as the difference between the current of the irradiated diode and the dark current. The dotted line depicts the dark current measured.

Figure 5. Photocurrent of figure 4 dependent on irradiation.

On the other hand, figure 6 shows the capacitance of the amorphous pin detector. This detector can well be described as a geometrical capacitance with an i-layer thickness, which describes the electrode distance. Because of the missing voltage dependence of the i-layer thickness, the capacitance varies only by $0.2\% \cdot V^{-1}$ in the relevant frequency range.

In the TFA technology, both capacitance dependences have to be considered. Therefore, as shown in figure 7, the influence of the crystalline pn-junctions increases with decreasing voltage. This is more important for smaller pixel areas. Dependent on the pixel concept a compressed transfer curve is obtained if the photocurrent integrates the image sensor signal down, and the curve is decompressed if the photocurrent integrates the signal up.

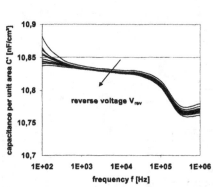

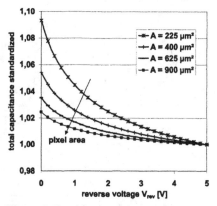

Figure 6. Capacitance of an amorphous pin-detector (from 0 V to -10 V in 1 V steps).

Figure 7. Total integration capacitance dependent on the reverse voltage, normalized to 5 V. The parameter describes different pixel areas.

DARK CURRENT OF A TFA IMAGE SENSOR

The dark current limits the dynamic range of an image sensor. For low light level applications, the sensor is cooled down to increase the dynamic range and sometimes to decrease the temporal noise. Therefore the sources of the dark current and its temperature behavior are important.

TFA image sensors have three dark current sources, which affect the integration capacitance: the leakage current of the detector diode, that of the reset transistor, and possibly that across the gate of the driver transistor. Figure 8 depicts the sources.

To estimate the influence of the amorphous and of the crystalline part, the following assumptions are made: the amorphous detector has a dark current density of 100 pA·cm^{-2}, the crystalline pn-junction one of 10 nA·cm^{-2}; the area of the crystalline diode is 0.25 µm^2, and the pixel area is 100 µm^2. Of the resulting dark current, a share of 80% is caused by the amorphous, and a share of already 20% by the crystalline silicon. This assumption includes not the subthreshold or many other leakage currents of the reset transistor. These currents can dominate the reset transistor leakage current, especially for deep submicrometer technologies [6], what results in an increasing dark current.

The gate leakage is important in connection with the further scaling down of the CMOS structure. 100 nm structures have a gate thickness of about 2.5 nm. This result in a leakage current density of $3 \cdot 10^{-3}$ A·cm^{-2} [7] with state-of-the-art materials. Using a source follower concept, the main part of this gate tunneling current flow to the drain. Compared with a dark current density of the amorphous detector described above and a very low gate area of 0.01 µm^2, the gate leakage current is 3000 times higher than that of the detector!

Both estimations show that for TFA sensors both the amorphous and the crystalline part should be considered. Especially with regard to scaling down the ASIC, the gate related dark current can have inadmissibly high values.

The temperature dependence of the pn-junctions in the crystalline ASIC is well known. It is caused by the density of intrinsic charges n_i and can be well described with the Sah / Noyce /

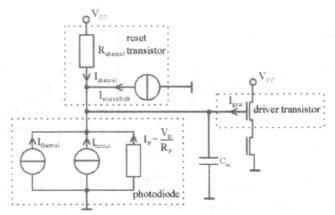

Figure 8. Dark current sources in a TFA image sensor with a source follower concept. The current directions depend on the TFA design configuration. Therefore some currents may be negative.

Shockley model. The dark current of the amorphous detector is more complex. For example, figure 9 shows the static dark current measurement of an amorphous pin diode with an i-layer thickness of 1.3 µm, an area of 36 mm^2, and at a measurement temperature of 50°C. Step width was 10 mV with 65 s recording time for each point.

This sample has three physical leakage current sources. At small reverse voltages the curve is dominated by the simple diode behavior J_{Sh} (see formula (6)). In the mid-range a parallel resistor R_{par} determines the curve (formula (7)), and at higher reverse voltages it results in the Poole-Frenkel effect J_{PF} (formula (8)), which is a bulk effect.

$$J_{Sh} \propto \exp\frac{q \cdot V}{n \cdot kT} - 1 \qquad (6)$$

$$J_R \propto \frac{1}{R_{par}} \qquad (7)$$

$$J_{PF} \propto V \cdot \exp\left(\frac{W_{bar} - q \cdot \sqrt{\dfrac{q \cdot E}{\pi \cdot \varepsilon_0 \cdot \varepsilon_{Si}}}}{kT}\right) \qquad (8)$$

In the above formulae, q, ε and k denote elementary charge, relative permittivity and the Boltzman constant, T describes the temperature, W_{bar} the energy barrier and E the electric field. The proportional factors are 95 pA·cm^{-2} for the reverse saturation current of the simple diode current, 1.5 GΩ·cm^{-2} for the parallel resistor R_{par} and 30 fA·cm^{-1}·V^{-1} for the Poole-Frenkel-Effect with a depletion thickness of 1.3 µm. The description of the dark current as the sum of (6), (7) and (8) is consistent with the measurements.

If only the intrinsic conduction density n_i is considered, the following formula is efficient to describe the temperature dependence:

$$n_i(T) \propto \left(\frac{T}{T_0}\right)^a \cdot \exp\left(-\frac{W_a}{kT}\right) \qquad (9)$$

The activation energy W_a and the temperature correction value a are dependent on the physical causes of the dark current. Amorphous pin diodes have a temperature correction value a in the range of one. This was described in a publication by Street [8] some years ago. Crystalline diodes have a value a in the range of 2 (for a space charge region thermal generation current) up to 4 (for a thermal generation current dominated on the minority carriers).

At room temperature, the exponential term in formula (9) dominates the temperature behavior. Therefore an Arrhenius plot results in straight lines (figure 10). In a first approximation, the measured values are about half of the band gap. This implies that the dangling bonds make an important contribution to the dark current behavior. Further, the curves have different slopes. This slope deviation describes different activation energies W_a. With increasing reverse voltage, the activation energy decreases (table I). However, despite the physical causes, formula (9) describes the temperature behavior sufficiently.

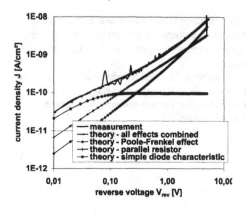

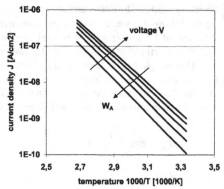

Figure 9. Dark current of an amorphous pin detector at a measurement temperature of 50°C, and its simple diode characteristic, parallel resistor and Poole-Frenkel effect shares.

Figure 10. Arrhenius plot of the dark current of an amorphous pin detector at a voltage of 1 V, 2 V, 3 V, 4 V, and 5 V.

For developing TFA sensors, a high activation energy has two advantages. First, it results in a low intrinsic conduction density n_i (formula (9)) and, therefore, in a low dark current. Second, with cooling down the sensor, the temperature difference is smaller. Table I shows that the sensor should have an operating point with a small reverse voltage, or the Poole-Frenkel effect should be negligible.

Reverse voltage [V]	1	2	3	4	5
Activation energy W_a [eV]	0.9	0.87	0.83	0.81	0.79
Temperature correction value a	1.2	1.4	1.3	1.1	1.1

Table I. Activation energy W_a and temperature correction value a for an amorphous pin diode with Poole-Frenkel effect at different reverse voltages.

TEMPORAL NOISE

Temporal as well as fixed pattern noise should be small so that a sufficient signal-to-noise ratio can be obtained. As yet, no concept describes the noise of the whole pixel input stage. While Blecher et al. [9] found an empirical description of the amorphous detector, this concept is not sufficient to explain the noise of the whole pixel input stage of a TFA image sensor. On the other hand, the development of crystalline ASICs is often done with simulation tools like SPICE.

A combination of these two concepts is very helpful to calculate the noise behavior of the whole pixel input stage. In this paper a concept is presented which transfers the measurement results of Blecher to a SPICE model.

In the Blecher model, the noise of the amorphous pin diode current has two parts: shot noise and flicker noise. Additionally, these parts have to be divided into a photocurrent and a dark current part. This distinction should be retained, because, while only the dark current noise is relevant to describe the dynamic range, both the photocurrent and the dark current noise are important to describe the signal-to-noise behavior.

The shot noise is proportional to the square root of the current and has a white spectrum. The flicker noise is more complex. Blecher described that this noise is related to the current by the following formula:

$$\overline{i^2_{x,1/f}} = \frac{c_x}{A^{2\beta_x-1}} \cdot I_x^{2\beta_x} \cdot \frac{1}{f^{\gamma_x}}$$ (10)

The appertaining values measured for the parameters in formula (10) are listed in table II.

I_x	c_x [cm^2]	γ_x	β_x
Short circuit photocurrent	$c_{photo} = 5 \cdot 10^{-16}$	$\gamma_{photo} = 1$	$\beta_{photo} = 0.85$
Reverse dark current	$c_{dark} = 2 \cdot 10^{-7}$	$\gamma_{dark} = 1$	$\beta_{dark} \approx 1$

Table II. Noise parameters to describe the flicker noise of an amorphous silicon pin diode.

In SPICE both noise sources are available in the diode model. The square of the shot noise current is directly proportional to the diode current, and the flicker noise can be described by the parameters AF and KF.

$$\overline{i^2_{1/f}} = KF \cdot I^{AF} \cdot \frac{1}{f}$$ (11)

Comparing formula (10) with formula (11), the SPICE parameter KF is dependent on the area of the amorphous diode and results in different values for the photocurrent and the dark current. Especially if the parameter γ equals one, a simple calculation can be used to determine the SPICE parameters. Figure 11 and Figure 12 show the good agreement between the measurement and the SPICE simulation.

However, if this model is utilized for noise calculations in the pixel input stage directly, the AC analysis small signal circuit generates incorrect results [10]. A more complex model has to be implemented. The noise is generated in a separate circuit and then coupled into the pixel circuit with noisefree current-current converters. Additionally, the differential resistor of the amorphous detector, which is noisefree, has to be substituted with a noisefree voltage-controlled current source. This concept is depicted in figure 13.

The noise is critical especially for a low sensitivity and, therefore, for a high input capacitance. So the following SPICE simulations are done with a pixel area of 0.0125 mm^2. For the input stage, a source follower concept with PMOS transistors (gate length 4 µm, gate width 4 µm) for the driver and the load are used. The SPICE transistor parameters are taken from the MOSIS AMIS 1.5 µm ABN process. Additionally, the BSIM3v3 noise model is used.

Figure 14 shows the spectral noise current dependent on the photocurrent. Until the photocurrent is below 100 pA, the noise of the pixel input stage is nearly independent of the detector noise. Figure 15 subdivides the noise causes. For low frequencies or, in other words, for long integration times, the flicker noise of the driver or the load transistor, respectively, domi-

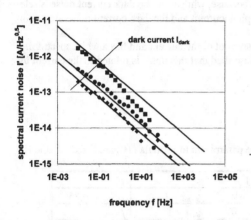

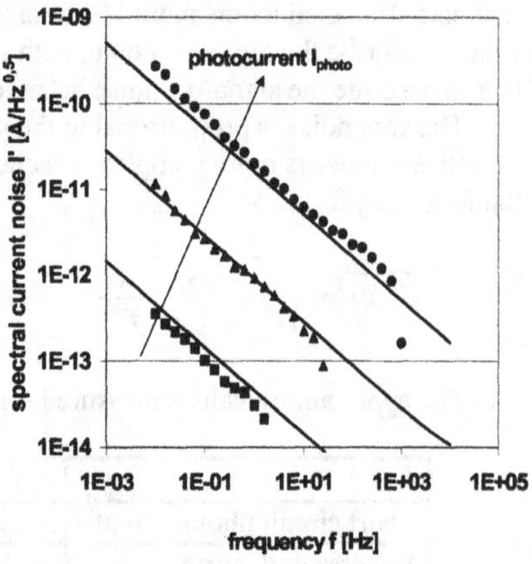

Figure 11. Flicker noise of the dark current, measurements and SPICE simulations for 5 pA, 9 pA, and 50 pA.

Figure 12. Flicker noise of the photo current, measurements and SPICE simulations for 30 nA, 1 µA, and 30 µA.

nates the total noise. In the mid range, the parallel resistor of the detector is dominant. If the photocurrent is high enough, the noise of this part has to be considered, too. For short integration times, the capacitance at the input node decreases the thermal noise of the resistor and the shot noise of the photocurrent. In this range the thermal noise of the transistor stage is the dominant noise source.

The comparison of different pixel input stages is significant. In figure 16 the noise behavior of all possible common source concepts (SC) as well as the source follower concepts (SF) with NMOS and with PMOS transistors are compared. To get comparable results all transistors had a

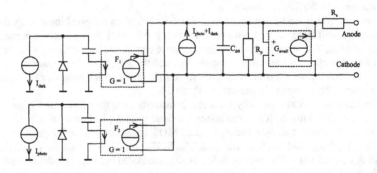

Figure 13. Concept which considers the dark and photocurrent noise in SPICE simulations.

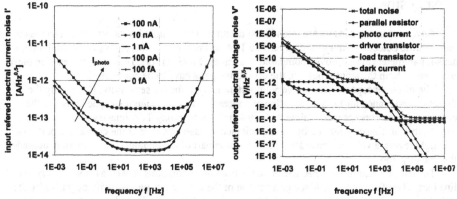

Figure 14. Noise of a PMOS source follower pixel concept. Noise is negligible for photocurrents lower than 100 pA.

Figure 15. Noise parts of a PMOS source follower concept and a photocurrent of 100 pA.

gate length / width of 4 μm / 4 μm. The lower 1/f noise of the PMOS transistors are caused by the older buried channel technology. However, this effect in not very significant for TFA image sensors as can be seen in figure 16. Because of the surface technologies for NMOS as well as for PMOS transistors in modern technologies, both transistor types have the same 1/f noise behavior.

The results of the PMOS/NMOS and the NMOS/PMOS common source circuits are conspicuous. These two curves show noise values that are higher by some orders of magnitude. Compared with the sensitivity described above, the noise increase is higher than the increase in sensitivity. Therefore, high sensitive image sensors may not be usable for low light applications. Low noise and high sensitivity may be conflicting goals.

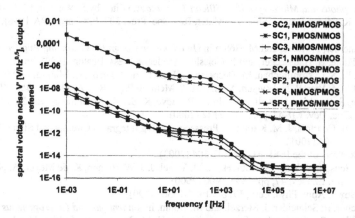

Figure 16. Noise of different Pixel input stages. SC means common source concept, SF source follower concept; the first name stands for the driver transistor type, and the second describes the load. More design parameters are listed in the text.

37

CONCLUSION

Advanced image sensors are intended to detect minimum light levels correspond to the circuit specific detection limit of the pixel. The TFA technology enables concepts in which image sensor parameters are improved compared to simple CMOS sensors. However, to describe these parameters, both the amorphous and the crystalline parts have to be considered.

The different capacitances are very important to describe the sensitivity. Using only the detector capacitance for the source follower concept or only the coupling capacitance for the common source concept is insufficient. Additionally, integrating TFA image sensors have a sensitivity non-linearity limited by the capacitance voltage behavior of the crystalline part. For low light levels, a voltage-dependent recombination part of the photogenerated carriers degrades the linearity, too.

The dark current of the TFA sensor stems from the amorphous as well as from the crystalline part. Therefore, the excellent characteristic of the amorphous part is partly degraded by the ASIC. Further, the activation energy of the amorphous detector is dependent on the operating point.

A concept is presented which implements the noise of the amorphous detector in a SPICE simulation. Resulting from the simulations with this concept, the dark current as well as the photocurrent noise can be neglected for low light levels. Further, input stages with high sensitivity and low noise can be conflicting goals.

Although the TFA technology enables better image sensors, the results presented in this paper can also be used to describe and optimize simple CMOS image sensors, neglecting the amorphous part.

REFERENCES

1. M. Böhm, F. Blecher, A. Eckhardt, K. Seibel, B. Schneider, J. Sterzel, S. Benthien, H. Keller, T. Lule, P. Rieve, M. Sommer, B. van Uffel, F. Liebrecht, R. C. Lind, L. Humm, U. Efron, E. Roth, in *Amorphous and Microcrystalline Silicon Technology*, edited by S. Wagner, M. Hack, H. M. Branz, R. Schropp, and I. Shimizu (Mater. Res. Soc. Proc. **507**, Pittsburgh, PA 1998), pp. 327-338
2. B. Schneider, P. Rieve, and M. Böhm in *Handbook on Computer Vision and Applications*, ed. by B. Jähne, H. Haussecker, and P. Geissler (Academic Press, Boston, 1999), pp. 237-270
3. G. Anelli, S. C. Commichau, M. Despeise, G. Dissertori, P. Jarron, C. Miazza, D. Moraes, A. Shah, G. M. Viertel, N. Wyrsch, Nucl. Inst. & Meth. In Phys. Res. A, **518**, pp. 366-372, 2004
4. T. Lule, S. Benthien, H. Keller, F. Mütze, P. Rieve, K. Seibel, M. Sommer, and M. Böhm, IEEE Trans. Elec. Dev. **ED-47**, pp. 2110-2122 (2000)
5. J. Goy, B. Courtois, J. M. Karam, F. Pressecq, Analog Integrated Circuits and Signal Processing, **29**, pp. 95-104 (2001)
6. H. Falk, Proc. of the IEEE **91**, pp. 303-304 (2003)
7. W. K. Henson, N. Yang, S. Kubicek, E. M. Vogel, J. J. Wortmann, K. de Meyer, A. Naem, IEEE Trans. El. Dev. 47, pp. 1393-1400 (2000)
8. R. A. Street, Appl. Phys. Lett., 57, pp. 1334-1336 (1990)
9. F. Blecher, B. Schneider, J. Sterzel, and M. Böhm, in *Amorphous and Heterogeneous Silicon Thin Films: Fundamentals to Devices*, edited by H. M. Branz, R. W. Collins, H. Okamoto, S. Guha, and R. Schropp, (Mater. Res. Soc. Proc. **557**, Pittsburgh, PA 1999), pp. 869-874
10. J. Sterzel, Bestimmung und Modellierung von Detektionsgrenzen bei TFA-Bildsensoren, thesis (2005)

Mater. Res. Soc. Symp. Proc. Vol. 869 © 2005 Materials Research Society

THIN-FILM COLOR SENSOR ARRAYS

D. Knipp[a,c], R.A. Street[a], H. Stiebig[b], M. Krause[b,d], J.-P Lu[a], S. Ready[a], J. Ho[a],

a.) Palo Alto Research Center, Electronic Materials Laboratory, Palo Alto, CA 94304

b.) Research Center Jülich, Institute of Photovoltaics, 52425 Jülich, Germany

c.) International University Bremen, Department of Science and Engineering, 28759 Bremen, Germany

d.) now with Infineon Technology, Dresden, Germany

ABSTRACT

Color information is commonly captured by silicon sensor arrays covered by a mosaic of color filters. However, the detection of the colors red, green and blue at different spatial positions of the sensor arrays leads to color aliasing or color moiré effects. This effect inherently limits conventional sensor arrays. In order to overcome this limitation we have realized color sensors based on vertically integrated thin-film structures. The complete color information can be detected at the same position of a sensor array without using optical filters. The sensors consist of a multilayer thin-film system based on amorphous silicon and its alloys. The spectral sensitivity of the sensors can be controlled by the optical and optoelectronic properties of the employed materials and the applied bias voltages. The working principle of the thin-film sensors and the sensor arrays will be presented. For the first time a large area three color sensor array was realized without using optical filters.

INTRODUCTION

Imaging is usually performed by silicon sensor arrays in combination with color filter arrays (CFA). The color filter array is typically realized by a spatial arrangement of at least three types of filters for the colors red, green and blue. Therefore, three chromatic color pixels are required to form a color pixel, which limits the resolution of conventional sensor arrays. Furthermore, color detection using CFAs leads to color moiré or color aliasing effects, which are observed when structures with high spatial frequencies are captured.

In order to overcome the color moiré effect, vertically integrated sensors have been proposed, which detect the color information in the depth of the sensor structure. Due to the wavelength dependent absorption of the semiconductor material, photons are absorbed at various depths, so that the color information can be detected in the depth of the device. Various sensor structures have been realized by using different materials, design concepts and contact configurations [1-9]. The suggested sensors range from two terminal devices, which change their spectral sensitivity by varying the applied bias voltage to vertically stacked diodes. So far only the sensor technology developed by Foveon has been commercialized [1]. The sensor structure consists of three vertical integrated pn-junctions fabricated by standard silicon BiCMOS processing. Sensor arrays with a resolution of 10.6Mpixels and metameric errors comparable to CFA based sensor arrays were presented [10]. Even though the color moiré effect is getting more pronounced for larger pixels, the realization of vertically integrated three color sensors for large area applications has not been demonstrated.

Large area sensor arrays are of interest for a variety of applications. In particular those applications are of interest, where the requirements in terms of cost and scalability are different from classical silicon microelectronics. Typical examples are biochips, neurochips and Lab-on chip systems. The dimensions of such systems are much larger than the features size of classical microelectronic components. Therefore, scaling of the device component is inherently limited. Typical examples are biochips or micro fluidic systems, where the dimensions of the micro fluidic channel are limited by surface tensions.

In this paper we present sensors, which consist of amorphous two terminal devices in combination with amorphous silicon thin-film transistor technology. The entire sensor array is realized on a glass substrate at low temperatures using a Plasma Enhanced Chemical Vapor Deposition (PECVD) process. After a brief description of the sensor concept and the fabrication of the sensor array, the experimental results of single sensors will be presented. In the second part of the paper the performance of the sensor array will be discussed.

EXPERIMENT

The sensor arrays were fabricated by plasma enhanced chemical vapor deposition (PECVD) at temperatures below 300°C on 4 inch glass wafers. A cross section of a sensor pixel of the sensor arrays is shown in figure 1. The optical sensors were integrated on top of the readout transistors to achieve a high area fill factor. The sensor array has a resolution of 512 x 512 pixels and a pixel pitch of 100μm x 100μm. Pixel addressing is realized by amorphous silicon thin-film transistors (TFTs). Each pixel contains a TFT, a storage capacitor, an address line, and a contact pad to the sensor occupying 67% of the pixel area. A pixel circuit of the sensor pixel is given in figure 2. The actual color sensor itself is realized by an anti serial connection of two amorphous silicon pin diodes. To planarize the readout transistors and to reduce the coupling between the optical sensor and the readout electronics a 3-10μm thick oxynitride layer was prepared on top of the readout transistor. Only the back contact and the n-layer of the optical sensor were patterned. The remaining layers of the sensor and the top electrode of the sensor array were unpatterned. A detailed description of the deposition parameters and the device performance of the amorphous silicon thin-film transistor are given elsewhere [11]. The top contact of the sensor was realized by radio frequency magnetron sputtered ZnO [12].

To minimize the color error of the sensor the spectral sensitivity of the detector has to be matched to the color space of the human eye represented by the colorimetric standard observer (CIE 1931). Therefore, bandgap engineering of the individual regions of the sensor is needed to optimize the spectral sensitivities. The bandgap diagram of the optical sensor is shown in figure 3. The top diode (the diode which is firstly penetrated by the incoming light) of the sensor is realized by a diode with an amorphous silicon carbon absorber. The absorber in the top diode has an optical band gap of 2.2eV.

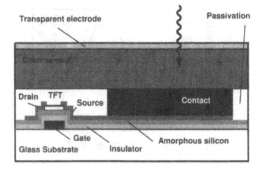

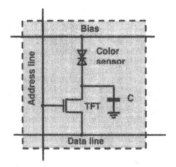

Fig. 1: Cross section of a sensor pixel of a large area color sensor array.

Fig. 2: Pixel circuit of a sensor pixel of a large area color sensor array.

Blue light is absorbed in the top diode, whereas green and red light is transmitted through the top diode. The absorber of the bottom diode is divided in two regions. Region i_{II} of the bottom diode was realized by a silicon carbon layer with an optical band gap of 2.0eV, whereas the region i_{III} was formed by an intrinsic amorphous silicon layer. A very thin lightly n-type doped layer was introduced between the two absorption layers of the bottom diode. The thin layer leads to an increased electric field in region i_{II} of the bottom diode. The spectral sensitivity of the sensor can be changed by varying the applied bias voltage. Applying a positive voltage to the nipin sensor leads to a reverse biased top diode and a forward biased bottom diode. The photocurrent is determined by the photogenerated carriers in the top diode, whereas the photogenerated carriers in the bottom diode recombine. Therefore, the detector yields blue sensitivity. Changing the applied bias towards a negative voltage leads to a forward biased top diode and a reverse biased bottom diode. In this case the sensor is green or green+red sensitive. For low negative voltages the electric field in region i_{II} is much higher than the electric field in region i_{III}, so that the photocurrent is determined by the photogenerated carriers in region II. For higher negative voltages the electric field is enhanced throughout the entire bottom diode and the photocurrent is determined by the extracted carriers out of region i_{II} and i_{III}. A detailed description of the deposition conditions of the sensor is given elsewhere [13].

RESULTS

The maximum of the voltage controlled spectral response of the nipin structure shifts from red to green and blue due to a change of the bias voltage from V=-3.5V to V=+1.5V. Figure 4 exhibits the spectral sensitivity of the sensor array for different bias voltages. For a bias voltage of +1.5V the sensor exhibits a maximum of the spectral sensitivity at a wavelength of 430nm. Applying a low negative bias of -0.6V to the optical sensor leads to a shift of the spectral sensitivity. Now the photocurrent is determined by the photogenerated carriers in region i_{II} of the bottom diode. The electric field in region i_{III} is not high enough to extract the photo-generated carriers out of this region. The sensor exhibits a maximum of the spectral response at 530nm.

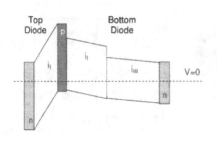

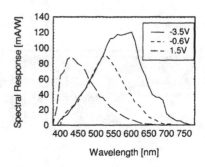

Wavelength [nm]

Fig. 3: Schematic band structure of the nipiin sensor under thermal equilibrium.

Fig. 4: Spectral response of the nipiin sensors for the applied bias voltages of V=+1.5V, -0.6V and -3.5V.

For higher negative bias almost all carriers can be extracted out of region i_{II} and i_{III} and the spectral response shifts to 600nm.

The pixel circuit of a sensor pixel is shown in figure 2. The charges created by the incoming light are stored by an external storage capacitor before being transferred to the readout circuit via the data line. The external capacitor is needed because the sensor structure does not allow for operation in the charge storage mode. Three images taken by the thin-film sensor array for different applied voltages are shown in figure 5. The images correspond to the raw data taken by the sensor array. An integration time of 50ms was used to read out each pixel. Due to the detection principle of the sensor 3 frames are needed to capture a complete color image. The images were taken for a photon flux of 100-1000lx. For intensities below 100lx the integration time of the sensor has to be increased. For low levels of illumination the sensor array performance is limited by the integration time rather than the readout time of the sensor array. The transient behavior of the sensor and the image lag of the sensor array are inversely proportional to the incident light intensity [8]. Real time imaging is not possible for low levels of light intensity.

All three images have a resolution of 185 x 280 pixels. The images were taken for different voltages of -3.5V -0.6V, and 1.5V. For low negative voltages (-0.6V) only the carriers were extracted from region i_{II} (figure 3) of the bottom diode (figure 5b). Accordingly the sensor is green sensitive. For high negative voltages all photogenerated carriers in the bottom diode were extracted, which means that the sensor is green+red sensitive (figure 5a). Due to a color transformation the signals can be separated in the red and green signal. The image in figure 5c was taken for an applied sensor voltage of 1.5V. In this case only carriers generated in the top diode were extracted. The contrast of the image taken for 1.5V is lower than the contrast of the two other images shown in figure 5a and 5b. This is caused by the fact that only the n-layer of the bottom diode and the back contact of the sensor were patterned. The remaining layers including the top diode were unpatterned. As a consequence the contrast of the blue image taken by the top diode is reduced and the pixel cross talk is increased. The results are confirmed by

Fig. 5a: Image taken for a voltage of -3.5V applied to the sensor.

Fig. 5b: Image taken for a voltage of -0.6V applied to the sensor.

Fig. 5c: Image taken for a voltage of +1.5V applied to the sensor.

measurements of the line spread and the modulation transfer function. The increased cross talk is caused by an electric field gradient in the p-layer of the sensor. Different levels of light intensity lead to different potentials in the p-layer. Therefore, a spatial variation of the light intensity leads to an electric field gradient along the p-layer, which causes pixel cross talk. Depending on the illumination conditions the potential in the p-layer can be changed by up to 150mV. A more detailed description of the device behavior under different illumination intensities is given elsewhere [14].

The metameric analyses of the spectral sensitivities in figure 4 reveals metameric errors comparable to standard color imagers. However, the analysis of the merged RGB color images exhibit increased color errors. Further work is needed to identify the limiting factors of the imaging system.

CONCLUSION

For the first time a moiré effect or color aliasing free large area sensor array was presented. The sensor array was realized on glass substrates with a resolution of 512 x 512 pixels. The optical sensor and the readout transistors of the sensor array were realized by amorphous silicon and its alloys at deposition temperatures below 300°C. The sensor array enables color moiré free color detection. Accordingly, the complete color information can be detected at the same position of a sensor array without using optical filters. The color detector based on the anti serial connection of two amorphous diodes. The sensors exhibit a good color separation and a high dynamic range. The developed large area sensor arrays are of particular interest for applications like biochips and Lab-on chips. For such kind of applications the requirements in terms of cost and scalability are different from classical microelectronics.

ACKNOWLEDGMENTS

The authors like to thank B. Rech, F. Finger, and their research groups from the Research Center Jülich for their support in preparing the color sensors and the team of the process line at the Palo Alto Research Center for providing the TFT backbones.

REFERENCES

1. R.B. Mirell, United States patent, Patent number: 5,965,865 (199).
2. P. Seitz, D. Leipold, J. Kramer, J.M. Raynor, SPIE Vol. 1900, p. 21 (1993).
3. D.P. Poenar, R.F. Wolfenbuttel, Appl. Opt. Vol. 36, No. 21, p. 5109 (1997).
4. T. Neidlinger, M. Schubert, G. Schmid, H. Brummack, Mat. Res. Soc. Symp. Proc. **420**, 147 (1996).
5. J. Zimmer, D. Knipp, H. Stiebig, H. Wagner, IEEE Trans. Electron Devices Vol. **45** No. 5, 884 (1999).
6. P. Rieve, M. Sommer, M. Wagner, K. Seibel, M. Böhm, J. Non Cryst. Solids **266-269**, 1168-1172 (2000).
7. H. Stiebig, J. Giehl, D. Knipp, P. Rieve, M. Böhm, Mat. Res. Soc. Symp. Proc. **377**, 517 (1995).
8. D. Knipp, H. Stiebig, J. Fölsch, F. Finger, H. Wagner, J. Appl. Phys. **83** (3), p. 1463 (1998).
9. F. Palma, Springer Series in Material Science Vol. **37**, R.A. Street (Ed.),, Springer-Verlag, Berlin, p.306-338 (2000).
10. R.F Lyon, P.M. Hubel, Proc. IS&T/SID Color Imaging Conference, 349(2002).
11. R.A. Street, Springer Series in Material Science Vol. **37**, R.A. Street (Ed.), Springer-Verlag, Berlin (2000).
12. O. Kluth, A. Löffl. S. Wieder, C. Beneking, L. Houben, B. Rech, H. Wagner, S. Waser, J.A. Selvan, H. Keppner, Proc. 26th IEEE PVSEC, pp. 715-718 (1997).
13. W. Luft, Y. Tuso, Hydrogenated amorphous silicon alloy deposition processes, Marcel Dekker, Inc., 1993.
14. B. Stannowski, H. Stiebig, D. Knipp, H. Wagner, J. Appl. Phys. Vol. **85** No. 7, 3904 (1999).

Mater. Res. Soc. Symp. Proc. Vol. 869 © 2005 Materials Research Society

Optimization of the metal/silicon ratio on nickel assisted crystallization of amorphous silicon

L. Pereira[1], M. Beckers[2], R.M.S. Martins[2], E. Fortunato[1], R. Martins[1]

[1]Departamento de Ciência dos Materiais, Faculdade de Ciências e Tecnologia, Universidade Nova de Lisboa and CEMOP, Campus da Caparica, 2829-516 Caparica, Portugal
[2]Institute of Ion Beam Physics and Materials Research, Forschungszentrum Rossendorf, P.O.B. 510119, 01314 Dresden, Germany

ABSTRACT

The aim of this work is to optimize the metal/silicon ratio on nickel metal induced crystallization of silicon. For this purpose amorphous silicon layers with 80, 125 and 220 nm thick were used on the top of which 0.5 nm of Ni was deposited and annealed during the required time to full crystallize the a-Si. The data show that the 80 nm a-Si layer reaches a crystalline fraction of 95.7% (as detected by spectroscopic ellipsometry) after annealed for only 2 hours. No significant structural improvement is detected by ellipsometry neither by XRD when annealing the films for longer times. However, on 125 nm thick samples, after annealing for 2 hours the crystalline fraction is only 59.7%, reaching a similar value to the one with 80 nm only after 5 hours, with a crystalline fraction of 92.2%. Here again no significant improvements were achieved by using longer annealing times. Finally, the 220 nm thick a-Si sample is completely crystallized only after 10 hours annealing. These data clear suggest that the crystallization of thicker a-Si layers requires thicker Ni films to be effective for short annealing times. A direct dependence of the crystallization time on the metal/silicon ratio was observed and estimated.

INTRODUCTION

For several years, polycrystalline silicon (poly-Si) received special attention aiming the use on large area devices, such as solar cells and thin film transistors for active matrix displays. One of the first techniques used to obtain poly-Si was by low-pressure chemical vapour deposition (LPCVD) that requires high process temperatures on the order of 600-650°C. However the demanding for low cost devices and consequent use of low cost substrates such as commercial glass, forced to a decrease on the poly-Si processing temperature [1,2]. Besides that, the crystallinity is not high enough for as deposited poly-Si and the material has much intra grain defects and high roughness [3]. So, lately, the production of poly-Si for device's application has been done by crystallization of amorphous silicon (a-Si), using either laser or thermal annealing. Solid Phase Crystallization (SPC) was the first technique employed to crystallize a-Si [4]. However, for this process the crystallization temperature is still too high and the time required for full crystallization is too long. Nowadays, excimer laser annealing (ELA) is widely adopted as crystallization method for low temperature poly-Si [5], but it has some problems such as non-uniform crystallization [1] and high production cost [6]. So, metal induced crystallization (MIC) emerged as an alternative crystallization technique overcoming these problems. Nickel (Ni), is known to dissolve in the a-Si weakening Si bonds and enhancing the nucleation of crystalline silicon at temperatures lower than 500 °C [7], below than its intrinsic crystallization temperature

(~600°C). The reaction occurs at the interlayer by diffusion and the driving force for the crystallization of a-Si is the difference on the free energy between amorphous and crystalline phases. Nevertheless, one critical limitation of MIC is the device's performance degradation due to the imbedded metal impurities (surface and bulk contamination). Most metals form deep dopants or act as recombination centres [8]. This is controlled by metal diffusion and solubility and it is related with the metal thickness deposited on the a-Si and the annealing temperature used. So, it is important to find a compromise between the metal and silicon thickness in order to reduce the metal contamination.

EXPERIMENTAL DETAILS

The a-Si samples with different thicknesses (80 nm, 125 nm and 220 nm) were deposited in a tubular LPCVD Tempress Omega Junior furnace on glass (Corning 1737) at 550°C, 40 Pa and with a silane (SiH$_4$) flow of 45 sccm. Before the metal deposition the samples were cleaned in buffered hydro-flouridric acid for 20 seconds in order to remove the native oxide and after that washed in de-ionized water. Ni layers with 0.2 and 0.5 nm were deposited over the a-Si samples by e-beam evaporation. The samples were then annealed for 2, 5, 10 and 20 h at 500 °C in an inert gas atmosphere. In order to evaluate the structural properties spectroscopic ellipsometry (SE) using a Jobin-Yvon ellipsometer and X-ray diffraction (XRD), using a Rigaku diffractometer, were employed. The crystalline volume fraction (f) was determined from SE data using a Bruggemann Effective Medium Approximation (BEMA) [9]. Coplanar aluminium contacts were evaporated for dark conductivity (σ_d) measurements. Rutherford Backscattering Spectroscopy (RBS) measurements were done to analyse the Ni depth profile inside the crystallized poly-Si layers.

DISCUSSION

The effect of the metal thickness on the crystallization of a-Si was studied using 0.5 nm of Ni deposited on silicon samples with 80 nm, 125 nm and 220 nm, which gives a metal/silicon ratio (R) of 0.0063, 0.004 and 0.0022, respectively. Fig. 1a) shows the XRD data of the thinner a-Si samples when annealed for 2, 5, 10 and 20h. The crystallization occurs preferentially through the <111> plans since these have lower formation energy [10]. The crystallization through these plans has been also observed by other authors [7, 8, 11]. Right after 2h annealing the crystallization is completed since no further improvements are observed on the <111> normalized peak intensity for longer annealing times. The SE data suggests also a full crystallization after only 2h with a crystalline fraction of 95.7% obtained using the BEMA model to simulate the experimental data (fig 1b). For 5, 10 and 20h the calculated crystalline fraction is 95.9, 96.8 and 97.8%, respectively. At this point it is important to refer that the crystalline fraction obtained by SE never reaches 100% since it is used three reference materials to simulate each layer that constitutes the model, namely the bulk and surface roughness. The three reference materials are poly-Si that simulates the grains, a-Si that simulates the reaming amorphous fraction and grain boundaries, and voids to simulate porosity and also grain boundaries [12,13,14]. Since the material is polycrystalline it has always grain boundaries and so some a-Si and voids will constitute each layer of the BEMA model. The activation energy of electrical conductivity (E_A) also shows that there are no changes for the different annealing times. The values remain around 0.5 eV as expected for good quality poly-Si [15].

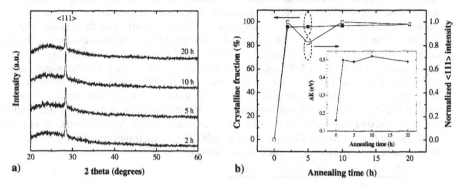

a) 2 theta (degrees) b) Annealing time (h)

Fig. 1 – Structural evolution of the 90 nm thick Si samples after annealed for different times: a) XRD diffraction data and b) Crystalline fraction determined by SE and normalized <111> peak intensity. The insert on fig 1b) shows the evolution of the E_A with the annealing time.

When using 0.5 nm of Ni to crystallize 125 nm of a-Si, a slight different behaviour is observed. The XRD and SE data (fig 2) show that for 2h annealing the crystallization is not complete. The relative peak intensity suggests a crystalline fraction around 65%, while the SE data gives a fraction of 59.7%. Only after 5h the crystallization seems to be complete, where <111> peak reaches its maximum intensity and the SE model points to a value of 92.2% for the crystalline fraction.

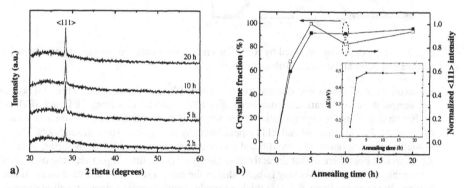

a) 2 theta (degrees) b) Annealing time (h)

Fig. 2 – Structural evolution of the 125 nm thick Si samples after annealed for different times: a) XRD diffraction data and b) Crystalline fraction determined by SE and normalized <111> peak intensity. The insert on fig 2b) shows the evolution of ΔE with the annealing time.

This value is improved up to 96.2% after 20h but neither the XRD data nor the electrical data (insert fig. 2 b) show any improvements.

As the crystallization occurs in a short time for thinner a-Si samples, it suggests that the metal/silicon ratio could be still reduced, in order to decrease the Ni incorporation inside the film. RBS data for crystallized Si samples with 80 and 125 nm are shown in fig 3. The results for longer annealing times are not presented since no differences are expected. It is clear the presence of a two-peak Ni region, meaning a non uniform Ni distribution throughout the Si. Integrating the overall area, the same values were obtained for all samples, confirming that the original Ni layer thickness was equal for all films. Calculating the channel difference in surface near Ni and at the end of Ni diffusion zone we obtain 729-679=50 for the 80 nm sample and 729-657=72 for the 125 nm sample. The ratio between them is 0.694, which is close to the ratio between 80 nm/125 nm=0.64 confirming that the Ni accumulation zones are located at the surface and at the substrate/silicon interface. Making the integration of the profile one reaches the metal/silicon ratios used. The minimal Ni concentration inside the films obtained was around 0.5% atomic percent for the 125 nm silicon sample. These data shows that the metal/silicon ratio could still be reduced since some metal saturation is observed.

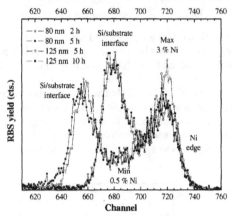

Fig. 3 – Ni depth profile obtained by RBS. Two metal accumulation regions appear one at the surface and other at the silicon/substrate interface.

Based on the RBS data it was decided to use 0.5 nm of Ni to crystallize a 220 nm a-Si layer. On this sample the crystallization is completed after 10h, as can be seen from XRD and SE data (fig. 4). For 5h annealing the crystallization is not yet completed, with a crystalline fraction estimated by SE of 81.7%. The normalized <111> peak intensity also gives a value around 80% The full crystallization was obtained after 10h and no enhancements were obtained by further annealing. An important point here is that the activation energy (E_A) is a little higher than the ones obtained for the thinner samples. This may indicate that as the ratio is reduced a decrease on the metal doping effect occurs. Expecting that thicker samples would require longer annealing times, it was decided do not analyse lower ratios since it seems that 10 h is an acceptable time limit for silicon crystallization. Fig. 5a shows the crystalline fraction obtained by SE for all samples. It is also plotted the structural evolution of a 80 nm sample crystallized with 0.2 nm of Ni, that gives a ratio of 0.0025. Fig 5b shows the full crystallization time as a function of the different metal/silicon ratios. As the ratio decreases the time needed to obtain full crystallization increases.

Despite having few points, when plotting the data on a logarithmic scale a linear dependence of the time on the ratio is obtained.

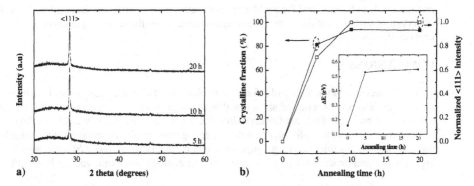

a) 2 theta (degrees)   b) Annealing time (h)

Fig. 4 – Structural evolution of 220 nm thick Si samples after different annealing times: a) XRD diffraction data and b) Crystalline fraction determined by SE and normalized <111> peak intensity. The insert on fig 4b) shows the evolution of E_A with the annealing time.

In the graph were plotted the points that represent the minimum full crystallization time obtained by extrapolation, using the intermediate crystallization state observed for ratios of 0.0040, 0.0025 and 0.0022 at 2 and 5 h. These values are, respectively 3.35, 5.50 and 6.15h. For the ratio of 0.0063 it was impossible to do this extrapolation since no intermediates points were determined.

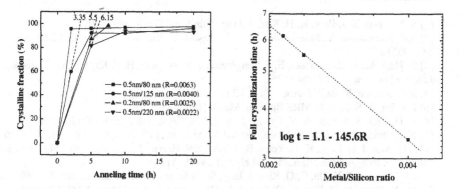

Anneling time (h)   Metal/Silicon ratio

Fig. 5 – a) crystalline fraction determined by SE for all samples and b) logarithmic relation between the metal/silicon ratio and the crystallization time.

Despite the error associated to this approximation, one can establish an empirical expression that relates the ratio with the annealing time on this range of metal/silicon ratios.

$$\log t = 1.1 - 145.6 \times R \tag{1}$$

A logarithmic approximation seems to fit more properly the experimental data since when extrapolating the full crystallization time for the ratio of 0.0062 a value around 1.5h is obtained while a linear relation would give a value around 0h, which is no credible. Of course, this relation needs to be analyzed for a wider range of ratios and for different metal thicknesses. An important contribution to this expression will also come from the annealing temperature that wa kept constant in this work.

CONCLUSIONS

This work focused on the ability of 0.5 nm Ni thick to crystallize a-Si samples with different thicknesses. The results achieved show that thinner samples with 80 nm corresponding to highe metal/silicon ratios (0.0062) are full crystallized at 500°C, for times shorter as 2h. As the ratio is reduced by increasing the a-Si thickness the annealing time required to crystallize the sample becomes higher. For this annealing temperature and on this range on metal /silicon ratios a logarithmic dependence was observed correlating the crystallization time and the Ni/Si ratio.

ACKNOWLEDGEMENTS

The author would like to thank to "Fundação para a Ciência e a Tecnologia" for the PhD scholarship. Apart from that, we would like to thank the financial support given by "Fundação para a Ciência e a Tecnologia" through pluriannual contract with CENIMAT. This work was performed in the frame of IRS Prime contract under the reference 03/00198 and "Fundação Lus Americana para o Desenvolvimento" (GSRT-CT-2001-05012) projects.

REFERENCES

[1] S. Y. Yoon, S. J. Park, K. H. Kim, J. Jang, Thin Solid Films, 383 (2001) 34
[2] S.-I. Muramatsu, Y. Minagawa, F. Oka, T. Sasaki, Y. Yazawa, Sol. Energy Mater. Sol. Cells 74 (2002) 275
[3] G. Harbeke, L. Krausbauer, E. F. Steigmeier, a. E. Widmer, H. F. Kappert, G. Neugebauer, J Electrochem. Soc., 131 (1984) 675
[4] R.B. Iverson, R. Reif, J. Appl. Phys. 62 (1987) 1675
[5] J. S. Im, R. S. Sposili, MRS Bulletin, March (1996) 39
[6] K. H. Kim, S. J. Park, A. Y. Kim, J. Jang, J. Non-Cryst. Solids 299-302 (2002) 83
[7] K. Andrade, J. Jang, B. Y. Moon, Journal of the Korean Physical Society, 39 (2001) 376
[8] J. H. Ahn, J. H. Eom, K. H. Yoon, B. T. Ahn, Sol. Energy Mater. Sol. Cells 74 (2002) 315
[9] D.A.G. Bruggemann, Ann. Phys. (Leipzig) 24 (1935) 636
[10] S. Y. Yoon, J. Y. Oh, C. O. Kim, J. Jang, Solid State Communications, 106, 6 (1998) 325
[11] S. Y. Yoon, K. H. Kim, C. O. Kim, J. Y. Oh, J. Jang, J. Appl. Phys., 82 (1997) 5865
[12] P. Petrik, T Lohner, M. Fried, L. P. Biró, N. Q. Khanh, J. Gyulai, W. Lehnert, C. Schneider H. Ryssel, J. Appl. Phys., 87 (2000) 1734
[13] L. Pereira, H. Águas, R. M. S. Martins, P. Vilarinho, E. Fortunato, R. Martins, Thin Solid Films, 451-452 (2004) 334
[14] L. Pereira, H. Águas, R. M. S. Martins, E. Fortunato, R. Martins, J. Non-Cryst. Solids 338-340 (2004) 178
[15] M. Hirose, M. Taniguchi, Y. Osaka, J. Appl. Phys., 50 (1979) 377-382

Optoelectronic Integration

Mater. Res. Soc. Symp. Proc. Vol. 869 © 2005 Materials Research Society   D1.5

Monolithic Integration of Electronics and Sub-wavelength Metal Optics in Deep Submicron CMOS Technology

Peter B. Catrysse
Edward L. Ginzton Laboratory, Department of Electrical Engineering, Stanford University,
Stanford, CA 94305-4088, U.S.A.

ABSTRACT

The structures that can be implemented and the materials that are used in complementary metal-oxide semiconductor (CMOS) integrated circuit (IC) technology are optimized for electronic performance. However, they are also suitable for manipulating and detecting optical signals. In this paper, we show that while CMOS scaling trends are motivated by improved electronic performance, they are also creating new opportunities for controlling and detecting optical signals at the nanometer scale. For example, in 90-nm CMOS technology the minimum feature size of metal interconnects reaches below 100 nm. This enables the design of nano-slits and nano-apertures that allow control of optical signals at sub-wavelength dimensions. The ability to engineer materials at the nanoscale even holds the promise of creating meta-materials with optical properties, which are unlike those found in the world around us. As an early example of the monolithic integration of electronics and sub-wavelength metal optics, we focus on integrated color pixels (ICPs), a novel color architecture for CMOS image sensors. Following the trend of increased integration in the field of CMOS image sensors, we recently integrated color-filtering capabilities inside image sensor pixels. Specifically, we demonstrated wavelength selectivity of sub-wavelength patterned metal layers in a 180-nm CMOS technology. To fulfill the promise of monolithic photonic integration and to design useful nanophotonic components, such as those employed in ICPs, we argue that analytical models capturing the underlying physical mechanisms of light-matter interaction are of utmost importance.

INTRODUCTION

Device dimensions in 21st-century complementary metal oxide semiconductor (CMOS) technology are reaching well into the nanometer regime [1]. While scaling enables increased clock speeds for central processing units (CPUs) and chip densities for random access memories (RAM), it also opens up opportunities for photonic device fabrication. In fact, integrated optics and photonics devices are already being implemented using processing techniques that are directly leveraged from semiconductor technology [2]. In this section, we briefly introduce CMOS technology and we evaluate some of its characteristics and its suitability from a photonics point of view.

CMOS technology is a complete process flow sequence, in which several technologies are combined to produce very large scale integrated (VLSI) circuit chips. Two important technologies in this flow are the front-end and the back-end technology.

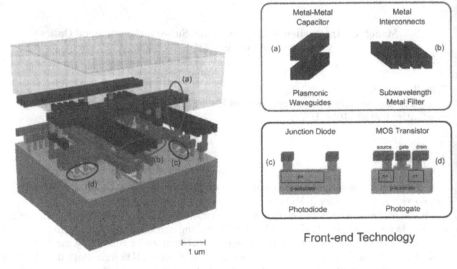

Back-end Technology

Metal-Metal Capacitor · Metal Interconnects · (a) · (b) · Plasmonic Waveguides · Subwavelength Metal Filter

Junction Diode · MOS Transistor · source gate drain · (c) · (d) · Photodiode · Photogate

Front-end Technology

1 um

Figure 1. Three-dimensional rendering of a 10 micron by 10 micron area of a VLSI chip implemented in CMOS technology. Shown is the silicon substrate (grayish-green surface at the bottom). Front-end technology defines the active devices, e.g., diodes (c) and transistors (d). The back-end technology defines capacitors (a) and interconnects (b). It consists of metal-1 (green rectangles), metal-2 (blue rectangles), and metal-3 wires (red rectangles) separated by dielectric layers (transparent) and connected by vias.

Front-end technology comprises a series of processing steps required to define active regions, N and P wells, gates, and sources/drains in the silicon (Si) substrate [3]. These steps are crucial for the creation of active devices (e.g., transistors or junction diodes), but are only introduced here for completeness. Here, we focus on the back-end technology, i.e., interconnect layers, contacts, vias, and dielectric layers that wire the active devices into circuits and systems.

While CMOS technology is a planar process, this does not imply that chips are two-dimensional. Figure 1 depicts a three-dimensional rendering of a 10 micron by 10 micron area of a VLSI chip. The insets identify some of the important structures that are created using back-end technology (e.g., metal-metal capacitors or metal interconnect wires). The back-end technology is responsible for creating most of the structures visible in this picture and typically stretches for about 10 micron above the Si substrate (grayish-green surface at the bottom). Low-resistance metal interconnects (green, red, and blue rectangles) wire the different parts of the chip and provide clock distribution. They are structured as a series of closely spaced parallel wires (red and blue rectangles). The interconnect layers are separated in the vertical dimension by high-resistivity dielectric layers. Vias (gray vertical stubs) connect two levels of interconnects. This completes the picture that a CMOS technologist has of the back-end technology.

However, one might pose the question: "What would a photonics researcher see when looking at the same picture?" Given the absorption properties of Si, the substrate is suitable for the detection of visible optical radiation. This can be achieved through the realization of photogates or photodiodes in front-end technology. These devices are nearly identical in implementation than the active devices used for electronic purposes (MOS transistors and junction diodes). In fact, photodetectors are currently being used as the photosensitive sites in the pixels of mass-produced charge-coupled device (CCD) and CMOS image sensors. In this paper, however, we focus on the optical properties associated with metallic structures that can be built in back-end technology. From a photonics point of view, an interconnect bus looks like a metal wire grid. The use of wire grids for filtering electromagnetic radiation goes back to the early experiments of Heinrich Hertz in the late 1800s [4]. In the field of microwave engineering, regular metal grids have been used for a long time as filters, transmitting some wavelengths and blocking others [5]. In fact, as a result of continued advances in the manufacturing technology of electronic ICs, metal grids with periodicity commensurate with the wavelength of near-infrared light have been demonstrated [6].

Over the past 25 years, the minimum lithographic feature size in CMOS technology has decreased by thirty percent every three years and has resulted roughly in a doubling of the number of transistors per chip every two years [1]. This trend is usually referred to as Moore's law [7]. Figure 2 shows the minimum half-pitch for a series of parallel wires implemented using the metal-1 layer in a CMOS logic process as a function of time [1]. While scaling is clearly being achieved to improve electronic performance of VLSI circuits, it also creates new ways to control light. With the minimum metal feature sizes available in 90-nm CMOS technology, sub-wavelength periodicities can now be achieved for the optical regime. This means that an interconnect bus or wire grid, for example, can be used as an optical filter to selectively change the transmission of visible electromagnetic radiation.

As device size decreases and chip density increases, interconnects have become a critical aspect of chip performance. Circuit time delays associated with interconnects have not kept pace with increasing device speeds. The resistance-capacitance (RC) time delay of a signal propagating along a metal interconnect (Fig. 3) is to first order obtained by treating it as a distributed, unterminated transmission line and is given by [3]

$$\tau_{\text{interconnect}} = 0.89\,RC\,, \tag{1}$$

where R is the line resistance and C is the total capacitance associated with the line. With scaling, interconnect width and separation follow the minimum feature size F_{min}. This means

$$\tau_{\text{interconnect}} \propto \frac{\varepsilon_{\text{dielec}}}{\sigma_{\text{metal}}} \frac{L^2}{F_{\text{min}}^2}\,, \tag{2}$$

where L is the interconnect length, while $\varepsilon_{\text{dielec}}$ and σ_{metal} represent the permittivity (or dielectric constant) of the dielectric surround and conductivity of the interconnect metal.

55

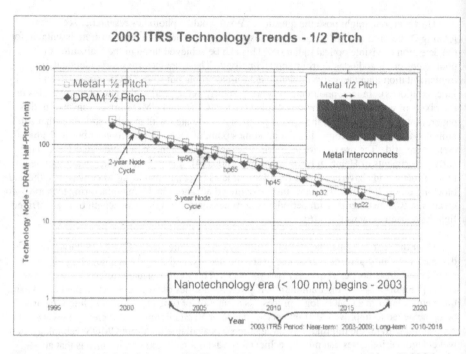

Figure 2. Scaling of the minimum pitch for metal interconnects implemented in a CMOS logic process. The figure shows the trend for both DRAM and metal-1 half-pitch according to the International Technology Roadmap for Semiconductors (ITRS) [1].

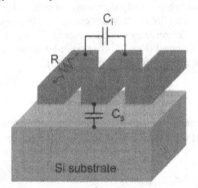

Figure 3. Interconnect structure for RC analysis. The green rectangles on top are the metal interconnects surrounded by SiO_2. At the bottom is the Si substrate (grayish-green). R is the line resistance, while C_i and C_s are the line-to-line and line-to-substrate capacitance, respectively. The total capacitance follows as $C = C_i + C_s$.

Global interconnect lengths usually increase as technology scales and thus RC delays tend to increase for a given interconnect material system. In order to maintain the distribution of clock and various other signals to different parts of a VLSI chip, however, RC delay time needs to keep pace with the decreasing gate delay time. This can be achieved by a careful choice of the materials.

The material system, on which back-end technology has traditionally relied, consists of aluminum (Al) for the metallic layers and silicon dioxide (SiO$_2$) for the dielectric layers. Aluminum features relatively high conductivity ($\sigma_{Al} = 3.64 \times 10^7\ S/m$), adheres well to SiO$_2$ and Si (due to the formation of its native oxide), and makes good electrical contact to heavily doped Si. SiO$_2$ has a relatively low static dielectric constant ($\varepsilon_{SiO_2} = 4.1$) and yields very controllable, stable and reproducible dielectric layers that serve to separate the metallic layers [3].

To reduce RC time delay in the face of scaling issues, one can increase conductivity and decrease permittivity. Three materials have higher conductivity than Al: silver, copper (Cu), and gold. Silver has corrosion problems and poor electromigration resistance, while gold has marginally higher conductivity than Al and suffers from device contamination issues. Cu has an almost-twice higher conductivity ($\sigma_{Cu} = 5.88 \times 10^7\ S/m$) than Al and has better electromigration properties. Hence, Cu has been chosen as a replacement for Al, even though this meant the introduction of a damascene process [3]. The dielectric constant (ε_{dielec}) of the interlayer dielectrics also affects device performance. To fulfill the requirements for VLSI, low dielectric constant (low-k) materials have been recently developed, e.g., hybrid organic-siloxane polymer (HOSP) with $\varepsilon_{HOSP} = 2.5$ [8] or Nanoglass with $\varepsilon_{Nanoglass} = 2.1$ [9]. This also results in a two-fold performance improvement.

While the back-end technology materials are chosen for their electronic properties, they are also suitable for doing photonics. For example, low-k dielectrics also have lower dielectric constants in the optical regime [8,9]. Since the wavelength of light λ is reduced inside the dielectrics ($\lambda = \lambda_0 / \sqrt{\varepsilon_{dielec}}$, where λ_0 is the wavelength in vacuum) that surround the metallic structures, a wire grid can only act as a zeroth-order diffraction grating when its period is smaller than λ. Therefore, reducing ε_{dielec} in the optical regime helps with scaling. Equally important are the losses introduced in the metal through Joule heating. It turns out that Cu not only has a larger static conductivity than Al, but also features lower losses in the optical regime [10,11].

In summary, the structures that are implemented in CMOS back-end technology feature both electronic and photonic functionality. The decrease in feature size, which is driven by an increase in electronic performance, also benefits photonic performance. Finally, the materials used in CMOS technology are chosen for their electronic characteristics, yet they also exhibit suitable photonic properties. These observations are the motivation for using of CMOS back-end technology as a platform for research in nanophotonics and light-matter interaction at the nanoscale with the potential for monolithic integration of both electronics and photonics.

MONOLITHIC INTEGRATION OF ELECTRONICS AND PHOTONICS

Electronic ICs have transformed the world we live in. They have propelled us into the information age and will continue to do so as CMOS technology keeps scaling. Current information technology also often uses light (photons) to transmit, store, and display information. Photonic technology is thus equally intimately interwoven with information technology and will continue to be so into the foreseeable future. It is therefore expected that monolithic integration of photonic and electronic ICs will not only advance information processing, but also will offer unique opportunities for future interconnects, telecommunication and biosensing devices.

To achieve monolithic integration, a major challenge needs to be overcome: in contrast to electrons, which have a very small footprint (<< 1 nm), visible and infra-red photons have sizable dimensions (100s nm). This inherent size mismatch aggravates the integration of electronic and photonic ICs. Recent advances in nanophotonics, i.e., the manipulation of light at a length scale smaller than its wavelength, place us in a unique position to start addressing this challenge. Nanophotonics includes both photonic crystals [12] and plasmonic devices [13]. Both approaches are central to molding the flow of light below the diffraction limit and, hence, to miniaturizing photonic ICs. It is therefore timely to explore the integration of photonic functionality with deep submicron CMOS technology.

In this section, we describe an early example of monolithic integration of electronics and sub-wavelength metal optics in deep submicron CMOS technology. We present experimental work from the field of CMOS image sensors [14], involving integrated color pixels, a novel color architecture for CMOS image sensors.

An important trend in digital camera design is the development of CMOS image sensors. These sensors are being scaled with CMOS technology to enable an increasing level of integration of capture and processing to reduce system power and cost [15]. Toward completing design integration of color image sensors in CMOS technology, we have explored the possibility of introducing wavelength selectivity with only standard processing steps, i.e., without inserting the traditional color filter array (CFA, Fig. 4a). Specifically, we have implemented filters that consist of sub-wavelength patterned metal layers placed within each pixel to control the transmission of light through the pixel to its photodetector. We refer to such pixel design as an integrated color pixel (ICP, Fig. 4b). ICPs may be useful for digital camera applications but may be even more suited to multispectral imaging and a variety of other image-sensing applications.

Monolithic integration of the optical filters inside image sensor pixels offers several potential advances. Placing the filters close to the photodetector reduces color cross talk, and pixel vignetting and increases the efficiency of microlenses. Controlling wavelength responsivity in the pixel design process eliminates the need for additional CFA manufacturing steps [14]. Why have wire grids not been used for color filtering in the visible light regime? To control light in the visible wavelength range with wire grids requires periodicities smaller than the visible wavelengths. This has been impossible until recent 180-nm CMOS technology and even with the wire sizes possible here, we are on the edge of controlling visible wavelengths.

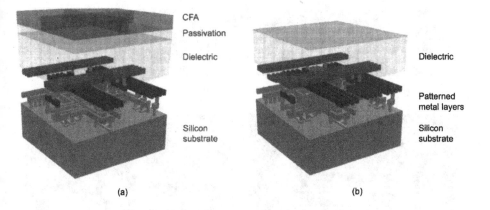

Figure 4. Geometry of four pixels in a CMOS color image sensor. (a) A conventional arrangement including a red-green-blue (RGB) color-filter array placed on the sensor surface. (b) The integrated color pixel arrangement that contains patterned metal layers (shown in color) within the pixel tunnels.

To investigate the use of sub-wavelength patterned metal layers as filters for color imaging, we implemented ICP test structures in a standard 180-nm CMOS technology. Periodic patterns were created in the metal layers already present in the process. These patterns comprised an array of elements whose gap widths were restricted by the process design rules and ranged from 270 to 540 nm. The refractive index of the surrounding dielectric (SiO_2) is 1.46. In air the visible wavelength regime is roughly 400-750 nm; in this medium the effective visible wavelength regime is roughly 270-510 nm. We implemented ICPs using a standard three-transistor active-pixel-sensor (APS) circuit. A photomicrograph of the chip with the ICP test structures is shown in Fig. 5a. The test structures (Fig. 5b) included one-dimensional (1D) and two-dimensional (2D) patterns with a periodicity (metal width + gap width) ranging from 540 to 810 nm and a gap width ranging from 270 to 540 nm.

The spectral transmittance through the patterned metal layers is measured indirectly by comparing the spectral responsivity of the ICP with that of an uncovered reference pixel. Figure 6 shows the spectral transmittance of a 1D patterned metal layer consisting of 270-nm-wide wires and gap widths ranging from 270 to 540 nm, where sizes represent the design specifications. Transmittance is shown for collimated illumination polarized with the electric field parallel to the wires (TE; Fig. 6a) and with the electric field perpendicular to the wires (TM; Fig. 6b). For TE polarization (Fig. 6a), for example, there is a constant but attenuated transmission region (left) and a cutoff region (right). The location of the cutoff region is determined by the gap width and the cutoff wavelength increases with increasing gap width. The TM transmittances in Fig. 6b differ in a few ways from the TE transmittances. Mainly, the TM transmittance measurements do decline at longer wavelengths with a steeper falloff which depends on the periodicity rather than gap width.

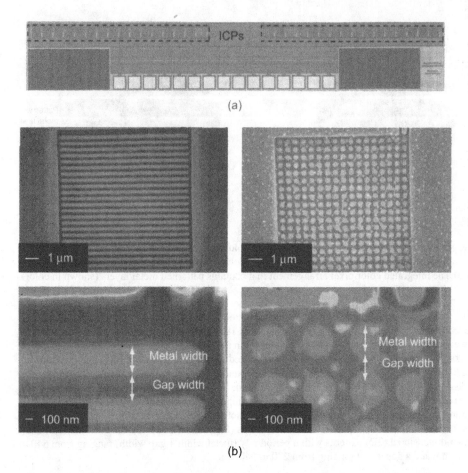

(a)

(b)

Figure 5. ICP implementation in 180-nm CMOS technology. (a) A photomicrograph of the ICP test structure chip. (b) Scanning electron micrographs of the ICP patterned metal layers. The 1D (left) and 2D (right) patterns are shown at two different spatial resolutions. The white spots, seen particularly on the right, are traces of material that was not removed completely during the deprocessing necessary to make these images.

Measured ICP transmittances are such that a linear combination, e.g., with a 3-by-3 matrix, suffices to produce red, green, and blue color channels with distinct peak sensitivities at approximately 750, 575, and 450 nm, respectively [16,17]. Although first-generation ICPs have a peak transmittance of only 40%, CMOS image sensors in 130-nm CMOS technology will permit smaller features and more flexibility in designing suitable patterns. This is expected to result in ICP transmittance comparable to that of CFAs.

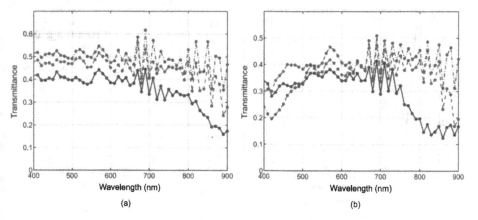

Figure 6. Measured transmittance of 1D ICPs with gap widths designed to be 270 (blue solid curve), 360 (green dashed curve), and 450 (red dash-dotted curve) nm. Wire width is held constant at 270 nm. Measurements are for collimated, polarized illumination: (a) E-field parallel with wires (TE) and (b) E-field perpendicular to wires (TM).

The measured TE and TM transmittance can also be predicted by using a numerical electromagnetic field simulation based on the finite-difference time-domain (FDTD) method [14]. FDTD is a first-principles method and is very valuable in analyzing existing structures. However, the method is slow and requires large amounts of memory, especially for 3D simulations. Moreover, the numerical aspect of the method does not provide the direct physical insight. This is very important when tackling an inverse problem such as filter design, i.e., determining the parameters of a structure so that it would generate the required spectral transmittance. To fulfill the promise of ICPs for color and multispectral imaging the design of patterned metal layers needs to be optimized and analytical models that capture the physical transmission mechanism are essential in guiding the design.

We have shown that a one-mode model [18], in which the sub-wavelength gaps of a 1D patterned metal layer are described in terms of single-mode waveguides, is in fact sufficient to capture the salient features of the ICP wavelength selectivity, even when the metal layer is modeled as a perfect electric conductor (PEC). The model provides closed-form Airy-like formulas for the transmittance of the patterned metal layers and gives physical insight into the transmission mechanism. Given the agreement with measurements, this analytical model should prove useful in optimizing the design of ICPs in CMOS image sensors and can also be straightforwardly extended to 2D patterns for polarization-independent color filtering.

Since feature sizes in 180-nm CMOS technology are not small enough to enable the implementation of the filters featuring the transmittance spectra shown in Fig. 7, we have developed an electron-beam lithography process. In this process, a combination of standard CMOS technology processing steps are combined with additional steps aimed at patterning a topical Al film (Fig. 8).

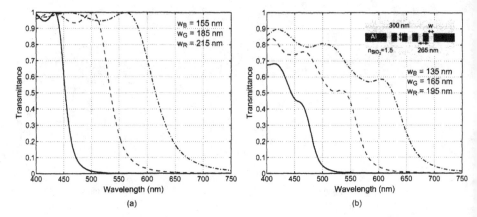

(a)   (b)

Figure 7. Transmittance of 1D sub-wavelength patterned metal layers suitable for RGB color filters in an ICP. The periodicity of the patterns (Λ) is chosen to assure sub-wavelength behavior for all wavelengths above 400 nm ($\Lambda = 265\,nm < 400\,nm/n_{SiO_2}$). (a) Transmittances obtained using the one-mode model in combination with a PEC assumption for the Al layer ($\lambda_{R,G,B} = 2n_{SiO_2}w_{R,G,B}$). (b) Transmittance obtained using 2D FDTD simulations with a Drude model for the Al layer. (R: red dash-dotted curve, G: green dashed curve, B: blue solid curve)

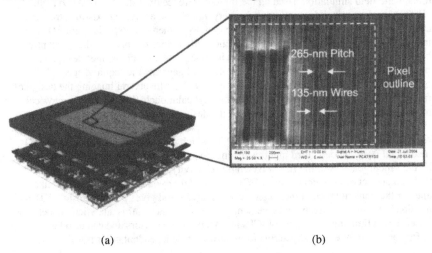

(a)   (b)

Figure 8. Electron-beam lithography process for sub-wavelength patterning of metal layers. (a) A topical Al layer is deposited on top of a processed 6-inch wafer with fully-functional monochrome CMOS image sensors. (b) Scanning electron micrograph of the sub-wavelength patterned photoresist used as a mask for the etching of the Al layer. Dimensions are for the blue filter with transmittance shown in Fig. 7.

Here, the Al film is applied at the surface as opposed to being embedded in the back-end technology. Yet, the electron-beam lithography process will enable us to define the sub-wavelength features required to prototype and evaluate the performance of the novel ICP color filters using existing 250-nm CMOS image sensor technology. This work is still in progress.

DISCUSSION

We explored the monolithic integration of photonic functionality with CMOS technology at the sub-wavelength scale. The structures that can be implemented and the materials that are used in CMOS IC technology are typically optimized for electronic performance. Upon closer examination, they are also suitable for manipulating and detecting optical signals. We showed that while CMOS scaling trends are motivated by improved electronic performance, they are also creating new opportunities for controlling and detecting optical signals at the nanometer scale. For example, in 90-nm CMOS technology the minimum feature size of metal interconnects reaches below 100 nm. This enables the design of nano-slits and nano-apertures that allow control of optical signals at sub-wavelength dimensions. The ability to engineer materials at the nanoscale even holds the promise of creating meta-materials with optical properties, which are unlike those found in the world around us. As an early example of the monolithic integration of electronics and sub-wavelength metal optics, we focused on integrated color pixels (ICPs), a novel color architecture for CMOS image sensors. Following the trend of increased integration in the field of CMOS image sensors, we recently integrated color-filtering capabilities inside image sensor pixels. Specifically, we demonstrated wavelength selectivity of sub-wavelength patterned metal layers in a 180-nm CMOS technology. To fulfill the promise of monolithic photonic integration and to design useful nanophotonic components, such as those employed in ICPs, we argued that analytical models capturing the underlying physical mechanisms of light-matter interaction are of utmost importance.

ACKNOWLEDGMENTS

The author was supported by Hewlett-Packard and Philips Semiconductor fellowships through the Center of Integrated Systems (CIS) at Stanford University and by Agilent Technologies through a research grant from the Stanford Photonics Research Center (SPRC). Brian A. Wandell and Abbas El Gamal provided valuable guidance during the CMOS image sensor work. Boyd Fowler, Michael Godfrey, and Khaled Salama assisted in the design and tape-out of the ICP test chip. Jeremy Theil (Agilent Technologies, Inc.) and James Conway (Stanford Nanofabrication Facility) were instrumental in pushing forward the electron-beam lithography project.

REFERENCES

1.   ITRS, *International Roadmap for Semiconductors, Executive Summary* (2004).
2.   M. Loncar, T. Doll, J. Vuckovic et al., *Journal Of Lightwave Technology* **18** (10), 1402 (2000).

3. James D. Plummer, Michael D. Deal, and Peter B. Griffin, *Silicon VLSI technology: fundamentals, practice and modeling*. (Prentice Hall, Upper Saddle River, NJ, 2000).
4. Heinrich Hertz, *Annalen der Physik und Chemie* **36** (4), 769 (1889).
5. P.J. Bliek, L.C. Botten, R. Deleuil et al., *IEEE Transactions on Microwave Theory and Techniques* **28** (10), 1119 (1980).
6. H. Tamada, T. Doumuki, T. Yamaguchi et al., *Optics Letters* **22** (6), 419 (1997).
7. G.E. Moore, *Electronics* **38** (8), 114 (1965).
8. K. Postava, T. Yamaguchi, and T. Nakano, *Optics Express* **9** (3), 141 (2001).
9. K. Postava, T. Yamaguchi, and M. Horie, *Applied Physics Letters* **79** (14), 2231 (2001).
10. M. A. Ordal, R. J. Bell, R. W. Alexander, Jr. et al., *Applied Optics* **24** (24), 4493 (1985).
11. M. A. Ordal, L. L. Long, R. J. Bell et al., *Applied Optics* **22** (7), 1099 (1983).
12. J. D. Joannopoulos, P. R. Villeneuve, and S. H. Fan, *Nature* **386** (6621), 143 (1997).
13. W. L. Barnes, A. Dereux, and T. W. Ebbesen, *Nature* **424** (6950), 824 (2003).
14. P. B. Catrysse and B. A. Wandell, *Journal Of The Optical Society Of America A* **20** (12), 2293 (2003).
15. A. El Gamal, David Yang, and Boyd Fowler, in *Sensors, Cameras, and Applications for Digital Photography*, edited by Nitin Sampat and Thomas Yeh (SPIE, Bellingham, 1999), Vol. 3650, pp. 2.
16. P.B. Catrysse, B.A. Wandell, and A. El Gamal, in *2001 International Electron Devices Meeting - Technical Digest* (IEEE, Piscataway, NJ, 2001), pp. 559.
17. Brian A. Wandell, *Foundations of vision*. (Sinauer Associates, Sunderland, Mass., 1995).
18. P. B. Catrysse, W. J. Suh, S. H. Fan et al., *Optics Letters* **29** (9), 974 (2004).

Infra-Red Photo-Detectors Monolithically Integrated with Silicon-Based Photonic Circuits

Jonathan D B Bradley, Paul E Jessop and Andrew P Knights

Department of Engineering Physics, McMaster University, 1280 Main Street West, Hamilton, Ontario, L8S 4L7, Canada.

ABSTRACT

The development of monolithic silicon photonic systems has been the subject of intense research over the last decade. In addition to passive waveguiding structures suitable for DWDM applications, integration of electrical and optical functionality has yielded devices with the ability to dynamically attenuate, switch and modulate optical signals. Despite this significant progress, much higher levels of integration and increased functionality are required if silicon is to dominate as a substrate for photonic circuit fabrication as it does in the microelectronic industry. In particular, there exists a requirement for efficient silicon-based optical sources and detectors which are compatible with wavelengths of 1.3 and 1.5µm. While a great deal of work has focussed on the development of silicon-based optical sources, there has been less concentrated effort on the development of a simple, easily integrated detector technology. We describe here the design, fabrication and characterization of a wholly monolithic silicon waveguide optical detector, utilizing an integrated p^+-v-n^+ diode, which has significant response to optical signals at the communication wavelength of 1.54µm. Measurable infra-red response is induced via the controlled introduction of mid-gap electronic levels within the rib waveguide. This approach is completely compatible with ULSI fabrication. The requirement for the detectors to be integrated with a rib waveguide and hence the guarantee of a long optical signal-device interaction, results in electrical signals of several µAs, even for deep-levels with a small optical absorption cross-section. Further, the rise and fall time of the detectors is compatible with current monolithic, silicon device based, optical switching and modulation operating in the MHz regime. These results suggest that these detectors offer a cost-effective route to signal monitoring in integrated photonic circuits.

INTRODUCTION

The advantages of silicon as a base material for the manufacture of integrated optical circuits (IOCs) have been well established by a number of authors [1]. Silicon is a low cost material which is virtually transparent at the important communication wavelengths around 1550nm, it has a relatively high refractive index which allows for the fabrication of compact device geometries, and it has excellent and well-understood electrical properties which allow the seamless integration of electrical and optical functionality on the same chip. Of greatest significance is the well-established and vast infrastructure which has been built upon many decades of research and high-volume production for the microelectronics industry. To date, a large range of silicon photonic devices has been demonstrated including low-loss waveguides, optical attenuators and (de)multiplexers, with a large on-going research effort aimed towards the development of efficient silicon-based optical sources [2]. Of some significance was the fabrication recently of an optical switch capable of operation in the GHz regime [3] and the demonstration of a silicon waveguide laser [4].

A considerable amount of recent work has been dedicated towards the development of integrated optical detectors, sensitive to wavelengths around 1550nm, the fabrication of which is completely compatible with standard silicon (CMOS) processing technology. This includes the development of Ge/Si heterostructures [5], the incorporation of optical dopants [6], such as erbium, into the silicon matrix, and the introduction of mid-bandgap energy levels via defect incorporation into the silicon lattice [7].

In this paper we describe a monolithically integrated p^+-v-n^+ silicon waveguide detector device structure compatible with current CMOS processing technology. Sensitivity to a broadband optical source with a wavelength range of 1530-1610nm is achieved via defect mediated carrier generation. The defects are introduced via proton bombardment of the waveguide structure and thus can be controlled in concentration, depth and location via selective masking, while shallow implantation and rapid thermal annealing of the p^+ and n^+-contact regions allows for the separation of the contacts to be close to the width of the waveguide, without introducing excessive free carrier absorption.

EXPERIMENTAL DETAILS

Device Fabrication

Low-loss optical waveguides in the rib geometry were fabricated using a silicon-on-insulator (SOI) substrate cut from a six inch wafer consisting of a low-doped ($<10^{15}$cm^{-3}) n-type 5μm thick silicon over-layer on a 1μm thick buried oxide, prepared using the bond-and-etch-back technique. A waveguide pattern of 11 individual ribs was photolithographically defined and etched into the silicon over-layer using a KOH solution. The rib widths of nominally 3μm at their base, and the rib heights measured to range from 0.9 to 1.1μm, ensured the waveguides were single mode at 1550nm. N-type and p-type doped regions were symmetrically defined on either side of the waveguide by masked phosphorus and boron ion implantation, respectively.

The separation between the doped regions (x) and the length of the doped regions (l) were varied for each waveguide. The dopant was activated using a 10 second anneal at 940°C in

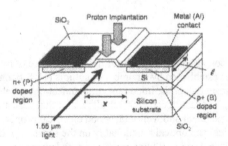

Figure 1. Illustration of monolithically integrated in-line power monitor for 1550nm in SOI, showing the junction separation (x) and device length (l).

flowing nitrogen, resulting in the retention of the dopant within a distance of 0.3μm from the surface. During annealing, the samples were capped using a PECVD deposited SiO_2 layer of 200nm thickness, through which contact vias were subsequently defined before the deposition and selective lift-off of an aluminum metallization layer. The final photolithography step defined areas for proton implantation that were centered on the rib, 50μm in width and of length l as defined by the doped regions on either side of the waveguide. The proton energy and dose were 170keV (equivalent to a range of 1.7μm in silicon) and $2 \times 10^{14} cm^{-2}$, respectively. Finally, optical quality waveguide end facets were prepared using a Loadpoint Microace dicing saw. This procedure does not require subsequent polishing or post-dicing facet preparation of any kind and has been found to result in facets of quality comparable with those produced via cleaving or dicing and polishing. The finished device structure is illustrated in figure 1.

RESULTS

Electrical Characterisation

For each waveguide, both current voltage (IV) and capacitance voltage (CV) measurements (at a measurement frequency of 1MHz) were performed.

Figure 2 shows a plot of IV for the device with x=9μm and l=6mm. Prior to the introduction of proton irradiation induced defects the diode exhibits a well-defined turn-on at 0.6V and a forward resistance of approximately 50Ω. A significantly increased forward resistance of around 300Ω, with a greatly softened turn-on, is observed in the implanted samples. This change in the IV curve may be attributed to the mechanism of Shockley-Read-Hall (SRH) recombination at the defect centers, which we assume here to be dominated by point defects such as the silicon divacancy. Figure 3 shows the CV relationship for the proton irradiated devices with device length l variation from 2 to 12mm, and diode contact separation x of 15μm. As expected, the capacitance increases as a linear function of device length and also decreases with increasing reverse bias. For a moderately reversed biased device of 2mm length, capacitance can be decreased to a value less than 30pF.

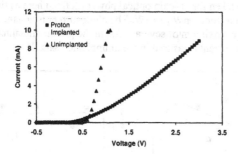

Figure 2. I-V characteristics for a proton implanted and an unimplanted device (with x = 9μm and l = 6mm).

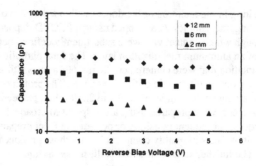

Figure 3. C-V characteristics for proton implanted devices (with $x =$ 15μm and $l = 2$, 6 and 12mm).

Optical Characterisation

Using a low loss, tapered optical fiber to butt couple light into each rib waveguide and a lens and free-space optical power meter at the output, the total power loss in each waveguide was determined. Figure 4a shows the measured waveguide loss plotted as a function of diode length l. A small but statistically significant increase in loss can be distinguished as l increases, which we estimate to be approximately 1.5dBcm^{-1}. This loss is attributed to the introduction of lattice point-defects, via the proton irradiation, which are known to increase optical attenuation in silicon [8]. Using the same data set we estimate an average fiber-to-chip coupling loss of 7±1dB. No significant variation in optical loss could be determined for a variation in x, confirming the insignificant absorption of the n^+ and p^+ contact regions, results reproduced in figure 4b.

Photoresponse

Figure 5 shows a plot of unbiased photocurrent in response to the waveguide coupled, broadband optical source observed for both an implanted and a non-implanted sample (l=6mm and x=9μm). Values were taken for on-chip optical power (defined here as the power coupled into the waveguide) ranging from 2mW to 1μW. The photocurrent is enhanced by more than an order of magnitude, up to a maximum of several μAs, in the proton implanted sample compared to the small intrinsic photocurrent measured in the unimplanted sample.

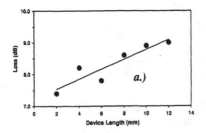 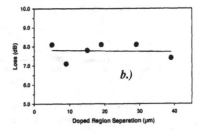

Figure 4. Measurement of optical loss versus a.)- diode length (l), and b.)- diode contact separation (x).

This enhancement is a direct result of carrier excitation by SRH generation via the defects introduced by the proton implantation. Further, the dark current was determined to be <1nA (the limit of sensitivity of our experimental set-up), suggesting a potential dynamic range beyond that shown in figure 5.

The measured photocurrent was observed to vary linearly with device length l. The photocurrent was also a strong function of separation (x) of the heavily doped regions. Assuming a mid-gap defect concentration of approximately $5 \times 10^{17} cm^{-3}$, consistent with the change in the IV curve for the implanted diode and previously measured concentrations of vacancy-type defects in proton irradiated silicon, the minority carrier lifetime was estimated to be on the order of nano-seconds, with a diffusion length <1µm. This results in the rapid recombination of the majority of photo-generated carriers. We conclude that the photocurrent will depend primarily on the overlap of the optical mode and the built-in potentials adjacent to the heavily doped n^+ and p^+ regions, where all generated carriers are assumed to be swept out by the built-in electric field. The most efficient device (for $x = 5$µm and $l = 6$mm) was measured to generate a photocurrent of 9.3µA, for approximately 2.7mW of on-chip optical power, giving a conservative estimate for responsivity of 3mA/W.

Frequency Response

The frequency response of the device with x=5µm and l=6mm was determined using an externally modulated laser source operating at a wavelength of 1550nm. The detector output was converted to a voltage of typically 150mV for an input power of 1mW using a transimpedence amplifier. The detector signal was monitored as a function of source frequency and the peak to value difference determined. Figure 6 shows this frequency response which indicates a 3dB bandwidth of 2.5MHz when unbiased, and 2.8MHz when the detector is reversed biased at 2V. This is compatible with current monolithic, silicon based, optical switches and modulators operating in the MHz regime [9].

SUMMARY

A fully CMOS-process compatible waveguide photodetector useful for optical power monitoring at 1550nm has been demonstrated. The infrared absorption properties of the silicon waveguide

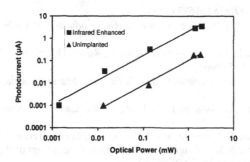

Figure 5. Photocurrent as a function of on-chip optical power for a proton implanted and an unimplanted device of dimensions $x = 9$µm and $l = 6$mm.

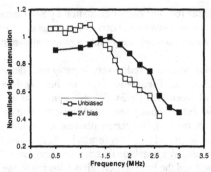

Figure 6. Frequency response of device with x=5μm and l=6mm.

are enhanced via defect mediated SRH carrier generation. Preliminary results show a responsivity of 3mA/W. The signal strength and absorbed optical power can be engineered by varying device geometries, location of defects and the defect density. Of significance is the potential for straightforward integration of this device with silicon-based photonic and electronic functionality (for example silicon waveguides and operational amplifiers). The bandwidth of the detectors is compatible with current monolithic, silicon based, optical switches and modulators operating in the MHz regime.

ACKNOWLEDGEMENTS

We would like to thank Ian Mitchell and Jack Hendriks at Interface Science Western, University of Western Ontario for invaluable assistance with development and implementation of the ion implantation process. We are also grateful to Jon Doylend and Phil Foster for useful discussion. This work is supported by the Canadian Institute for Photonic Innovation, the Natural Sciences and Engineering Research Council of Canada and the Ontario Photonics Consortium.

REFERENCES

[1] R. A. Soref, *Proc. IEEE*, **81**, 1687 (1993).
[2] G. Reed and A. P. Knights, *Silicon Photonics-An Introduction,* (John Wiley & Sons, Chichester, 2004).
[3] A. S. Liu, R. Jones, L. Liao, D. Samara-Rubio, D. Rubin, O. Cohen, R. Nicolaescu, and M. Paniccia, *Nature,* **427**, 615 (2004).
[4] H Rong, R Jones, A Liu, O Cohen, D Hak, A Fang and Mario Paniccia, *Nature*, **433**, 725 (2005).
[5] L. Colace, G. Masini, G. Assanto, H.-C. Luan, K. Wada, and L. C. Kimerling, Appl. Phys. Lett. **76**, 1231 (2000).
[6] P.G. Kik, A. Polman, S. Libertino, and S. Coffa, J. Lightwave Technol. **20**, 862 (2002).
[7] A Knights, A House, R MacNaughton and F Hopper, Proceedings of the Optical Fiber Communications Conference 2003 (OFC2003), 705.
[8] H. Y. Fan and A. K. Ramdas, J. Appl. Phys. **30**, 1127 (1959).
[9] W. Lin and T. Smith, *Kotura White paper on SOEICs*, http://www.kotura.com/.

Mater. Res. Soc. Symp. Proc. Vol. 869 © 2005 Materials Research Society   D4.4

Efficient Focusing with an Ultra-Low Effective-Index Lens Based on Photonic Crystals

Eugen Foca[1,2], Helmut Föll[1], Frank Daschner[3], Vladimir V. Sergentu[2], Jürgen Carstensen[1], Reinhard Knöchel[3], Ion M. Tiginyanu[2]

[1] Chair for General Materials Science, Faculty of Engineering, Christian-Albrechts-University of Kiel, Kiel, Germany;
[2] Institute of Applied Physics, Technical University of Moldova, Chisinau, Moldova;
[3] Microwave Laboratory, Faculty of Engineering, Christian-Albrechts-University of Kiel, Kiel, Germany.

ABSTRACT

This work focuses on photonic crystals (PC) that can be ascribed an effective index of refraction < 1 or even < 0. We investigate the possibility to design optical elements (in this case a lens) based on this type of PC. A new approach for determining the effective refractive index of PCs with unusual index of refraction is used, which is simpler than earlier methods based on analyzing equi-frequency surfaces in k-space. An ultra-low refractive index PC is given a form approximating a concave lens and is proven theoretically and experimentally that it efficiently focuses the electromagnetic radiation in the microwave range. Strong focusing effects are found for both polarizations (TE and TM mode). Intensity gains as large as 35 for TM polarizations and 29 for TE polarizations are found. Measurements are in a good accordance with simulations.

INTRODUCTION

Propagation of electromagnetic radiation in a Photonic Crystal (PC) is well understood in the long wavelength range, i.e. where the wavelength is large compared to the crystals' periodicity and energetically below the band gap. In this case PCs can be treated as homogeneous materials and all their parameters like n_{eff}, the effective index of refraction, are (relatively) easily calculated. In this case devices based on a PC have an $n_{eff} > 1$, given by $n_{eff} = ck/\omega$, where ω is the radiation frequency and k is the wave vector of the Bloch wave propagating in the PC.

However, more interesting phenomena are encountered in the short wavelength limit (energies above the band gap), where unusual dispersion functions and beam propagation are found. In this wavelength range, PCs may, for example behave as homogeneous material with an ultra-low index of refraction, meaning a $n_{eff} < 1$ or even a $n_{eff} < 0$ [1]. However, an index of refraction may also not be defined at all, or only in some approximation. We will therefore first discuss a method that allows to check rather easily if for a given wavelength and polarization state a given PC can be described, at least approximately, by an effective index of refraction, and what numerical value must be assigned to n_{eff} in this case. Using this method, suitable wavelength ranges are identified for which designing optical elements (here a concave lens) with ultra-low n_{eff} makes sense. The resulting lens is characterized exhaustively in two configurations: perfect order (perfect crystal) and strongly disturbed ("amorphous"). In what follows we report on extensions of the previous work [2] and present new results.

"PROBE MEDIUM" APPROACH

The probe medium approach exploits an earlier idea of Veselago [3]: At the interface of two media that have refractive indexes with identical absolute values, but different signs, no reflection will take place. As a result, a point source embedded in medium 1 at some distance to the interface, will be perfectly imaged in medium 2 on the other side of the interface (neglecting size effects); cf. Fig. 1a. Let medium 1 be a PC with an embedded point source, and now calculate the wave propagation for a homogeneous probe medium 2 with an adjustable index of refraction. Since the range of possible n_{eff} for the PC is limited, it is easy to see within a few relatively quick simulations, if point to point imaging is possible at all and if yes, which value of the index of refraction n_p of the probe medium produces the best image. Fig. 1b shows a case of optimal imaging; note that the small dimensions of the set-up cause the apparent non-perfection for this case. Larger PCs would give much better imaging, but this is not necessary at this stage and only increases computation time.

A result as shown in Fig. 1b thus asserts that i) an effective index of refraction can be defined for the case investigate, and ii) its value is given by $n_{eff} = -n_p$.

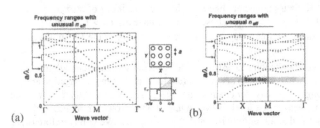

(a)   (b)

Figure 1 (a) Schematic view of the "probe medium" approach. (b) An example of simulation results for the case that n_{eff} exists.

PHOTONIC CRYSTALS WITH AN ULTRA-LOW INDEX OF REFRACTION

Using the probe medium approach the interesting (i.e. $n_{eff} < 1$) wavelengths regions for a simple PC consisting of dielectric rods with diameter d and dielectric constant ε_r arranged in a square lattice with lattice constant a were identified; the result is shown in Fig. 2 for the electrical or magnetic field of the incoming radiation transversal to the rods (TE and TM mode), respectively.

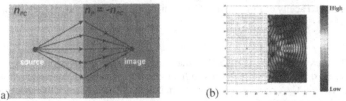

Figure 2 Dispersion diagram for the (a) transversal electric polarization and (b) transversal magnetic polarization. Intervals where $n_{eff} < 1$ are outlined for both polarizations.

(a)   (b)

The lattice constant of the PC is $a = 2.8$ cm and the radius of the cylinders is $d = 0.19a$. The dielectric constant was chosen as $\varepsilon_r = 9.6$, which gives a refractive index of $n_r = 3.1$, consistent with Al_2O_3 as material. As the dimension suggests, the corresponding PC is scaled to the microwave frequencies of $f \approx 10$ Ghz. It goes without saying, however, that the basic set-up scales linearly to any frequency; their features are identical to dispersion curves calculated with more involved methods. In quantitative terms, the effective index of refraction, calculated using the probe medium approach, will generally be smaller than unity for the following (approximate) values of a/λ: $a/\lambda \in (0.5;0.68) \cup (0.7;0.86) \cup (1;1.15)$ for the TE mode and $a/\lambda \in (0.67;0.92) \cup (1;1.1)$ for the TM mode. From this calculations it can be expected that a concave lens shaped from this PC will focus the radiation in the frequency regions outlined above. In what follows we prove that this is indeed the case.

EXPERIMENTAL RESULTS AND DISCUSSION

The concave lens used for the experiments consists of 112 alumina cylinders and has the form shown in Fig. 3a or 3b. While Fig. 3a shows a lens "cut" form a perfectly periodic PC, Fig. 3b shows an arrangement, where each rod has been randomly displaced from the regular lattice position by up to 30 %; this lens will be called "amorphous".

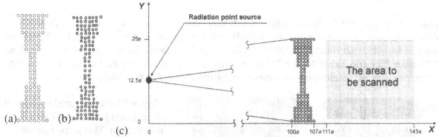

Figure 3 (a) Top view of the measured lens that consists of periodically arranged alumina rods. (b) An "amorphous" lens is used as well. c) The electromagnetic field distribution is scanned behind the lens at a height of half of the rod length.

All experiments are performed in an anechoic room that is "dressed" with special microwave absorbing plates. A radiation point source is placed at 100a from the lens edge (Fig. 3c). The field behind the lens is scanned in the (XY) plane that cuts the rods axis in the middle. All the scans involve an area of 32a x 25a, see Fig. 3c. The scanning resolution is 0.5 cm. Fig. 4 compares several experimental results with the respective simulations; including cases where n_{eff} is not defined. Shown is the power gain distribution that is calculated as the ration of the $E^2(x,y)/E_0^2(x,y)$, where $E^2(x,y)$ is the local electromagnetic power in the presence of the lens and $E_0^2(x,y)$ is the local electromagnetic power in its absence. The simulations are done using a somewhat modified version of the multi scattering approach [4,5]. Fig. 5a shows the case when focusing hardly exists (note the scale!). The measured power gain distribution is relatively noisy, but still in good agreement with the simulations. Fig. 4b shows a peculiar case for the TE polarization: Beam splitting, a potentially useful feature, indicating that in wavelength regions where obviously n_{eff} does not exist, PCs may still be used for special applications. While in this case both excited modes are relatively weak, they exist, and optimization appears possible. Finally, Fig. 4c and 4d give clear evidence of focusing for both polarizations. While efficient focusing was achieved, and the measured results agree quite well with the simulations, a quantitative comparison shows some minor discrepancies, which indicate that more work is needed for a complete understanding.

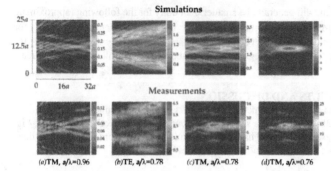

Simulations

Measurements

(a)TM, a/λ=0.96   (b)TE, a/λ=0.78   (c)TM, a/λ=0.78   (d)TM, a/λ=0.76

Figure 4 (a) – (d) Simulated and measured power gain of the concave lens for the TE and TM polarizations and various configurations. Note the difference in the power gain scale! For details refer to the text.

Fig. 4c-d indicate that the focusing lobe is still very "noisy", i.e. the focal spot is not clearly defined because of the slow radiation decay from the middle of the focusing spot. Fig. 5 presents the best focusing that could be obtained for the TE and TM mode. So far, power gains as high as 29 for the TE mode and 35 for the TM mode could be achieved. To the best of our knowledge no other experimental results were presented so far with comparable focusing power.

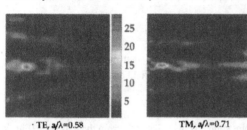

TE, a/λ=0.58   TM, a/λ=0.71

Figure 5 Best measured power gain distribution for TE and TM polarization. This is the highest efficiency that could be measured with the lens under investigation.

Nominally, the wavelength λ_{PC} of radiation propagating inside a PC with an ultra-low index is large since it is given by $\lambda_{PC} = \lambda_{vac}/ n_{eff}$ and one might expect that this large effective wavelength is not sensible to crystal irregularities. If that expectation can be transferred to optical elements like the concave lens that in itself is smaller then or just comparable to λ_{PC} is doubtful and best checked by experiments. Generally, it is difficult to calculate this effect since an amorphous structure like the one shown in Fig. 3b, will definitely not have any band structure like those shown in Fig. 2a-b.

While simulations predicted that a displacement of the cylinders up to 30 % from their regular positions would not fully destroy the focusing effect, this was only partially confirmed by the experiments. The amorphous lens usually does not show focusing in frequency regions where the regular lens focuses, but focusing was found in other regions.

In Fig. 6 and Fig. 7 a direct comparison of the intensity gain distribution for the regular lens and the amorphous lens is presented. For the TE polarization is observed that the amorphous structure "collects" better the rays. For some frequencies, where for the regular lens no evident focusing spot could be defined, the amorphous lens focused the radiation, however weakly, c.f. Fig. 6a and Fig. 6c. Note also that the intensity gain remains unaffected, although it is relatively small.

Why the amorphous lens focuses the radiation for some frequencies where the regular lens simply behaves as a strong disperse medium as is shown in Fig. 6b, is open to speculation at present. The focusing spot is poorly defined, and its intensity is relatively weak, but that the rays tend to converge to a spot is clearly visible.

Amorphous lens

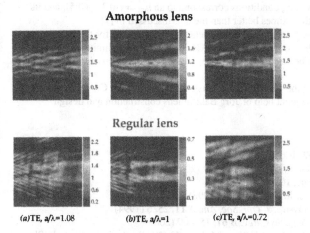

(a)TE, a/λ=1.08   (b)TE, a/λ=1   (c)TE, a/λ=0.72

Figure 6 (a) – (c) Experimental results for the TE polarization and various wavelengths in comparison to the field distribution for the crystalline lens.

Regular lens

In Fig. 7 several cases of focusing for the amorphous lens are shown for the TM polarization. As for TE polarization, the amorphous lens tends to focus better for the wavelengths used than the crystalline lens, c.f. Fig. 7a-b. As the wavelength increases, the focusing lobe becomes sharper, but the focusing power gets weaker. In contrast to the TM polarization the focusing spot area becomes smaller with increased wavelength. Thus, for $a/\lambda = 0.64$ the area of the focusing spot is about $0.65 \cdot \lambda^2$, which is well beyond the focusing limits of generic lenses known from linear optics. However, it is certainly premature to invoke the term superlens at this point.

Amorphous lens

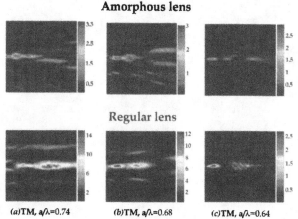

(a)TM, a/λ=0.74   (b)TM, a/λ=0.68   (c)TM, a/λ=0.64

Figure 7 (a)- (c) Focusing results for the amorphous lens in the case of TM polarization in comparison with the results obtained with the regular lens for increasing wavelength.

Regular lens

CONCLUSIONS

Some earlier works also discussed possible lenses made from PCs [6,7], but good focusing with unusual n_{eff} has not been shown before. Gupta and Ye [6], e.g., proposed and measured a convex lens based on alumina rods PC, but with an $n_{eff} > 1$. For our lens, having a radius of curvature of 62.5 cm, best focussing conditions correspond to an $n_{eff} = (-0.3 - +0.5)$, and its focusing efficiency is at least three times better than the lens discussed in [7].

In conclusion, our approach to unusual index metamaterials allowed for the first time to produce efficient "unusual index" devices based on PCs. This work is not limited to the PCs based on dielectric rods. It can be also extend to lenses done based on porous dielectrics [8] or 3D PCs.

The authors acknowledge the very helpful discussions with Prof. Dr. Georgi Popkirov and Dr. Sergiu Langa, as well as the great help of Jörg Bahr in lens construction and design.

REFERENCES

[1] M. Notomi, *Phys. Rev. B* **62**, 10696 (2000)

[2] E. Foca *et al.*, *Phys. Stat. Sol.* (a) 202, 4, R35-R37 (2005)

[3] V.G.Veselago, *Usp. Sov. Fiz.*, **10**, 509 (1968)

[4] L.-M. Li, Z.-Q. Zhang, *Phys. Rev. B* **58**, 9587 (1998)

[5] D. Felbacq, G. Tayeb, D. Maystre, *J. Opt. Soc. Am. A* **11**, 2526 (1994)

[6] Bikash C. Gupta and Zhen Ye, *Phys. Rev. B* **67**, 153109 (2003)

[7] Bikash C. Gupta and Zhen Ye, *J. Appl. Phys.* **94(4)**, 2173 (2003) ; Suxia Yang *et al.*, *Phys. Rev. Lett.* **93(2)**, 024301 (2004)

[8] V.V. Sergentu *et al.*, *Phys. Stat. Sol.* (a) 201, R31 (2004)

MBE Growth of GaAs on Si through Direct Ge Buffers

Xiaojun Yu[1,2] , Yu-Hsuan Kuo[2] , Junxian Fu[2], James S Harris, Jr.[2]
[1]Department of Materials Science and Engineering, Stanford University,CA 94305
[2]Solid State and Photonics Lab. Stanford University, CISX 126, 420 Via Ortega, Stanford, CA 94305

ABSTRACT

The result of GaAs growth on Si using a thin Ge buffer layer (about 0.5µm thick) is presented in this paper. A two-step method with a high temperature anneal between two steps is used to grow the Ge buffer layer. Single phase GaAs is grown on Ge by controlling the growth temperature, substrate miscut and the prelayers. No APD defect is observed by the XTEM and the threading dislocation density of GaAs grown using this method is about $5\sim10\times10^7 cm^{-2}$. The PL intensity of GaAs is 10× less on Si substrate than on GaAs substrates.

INTRODUCTION

The realization of low defect density GaAs on Si heteroepitaxy would enable monolithic integration of III-V materials and devices with conventional Si integrated-circuit (IC) technology. It is also desirable to grow high quality GaAs with a very thin layer for the application of high electron mobility transistors on Si. However, GaAs/Si heteroepitaxy has been a difficult challenge with insufficient success for two decades due to its two growth characteristics: large lattice mismatch (4%) and polar-on-nonpolar growth. The strain relaxation process causes the formation of dense threading dislocation. The polar-on-nonpolar growth results in the competition of two phases with different orientation, which creates antiphase defects and induces a rough surface [1]. Tremendous efforts have been made on the direct growth of GaAs on Si[2], and it is difficult to reduce the threading dislocation density (TDD) of the GaAs films to below $10^8 cm^{-2}$. Different techniques such as AlGaAs/GaAs superlattice buffer [3], strained superlattice and cycling anneal[4] have been used to reduce the threading dislocation but with the compromise of dramatically increase of the surface roughness. The graded Si_xGe_{1-x} buffer layer helps to reduce the TDD to below $10^7 cm^{-2}$, but a thick grading layer is required with the total thickness at $8\sim10µm$[5,6]. The thermal expansion mismatch between Si and epitaxial film is large and a thick SiGe buffers will cause problems during the following processing. The surface is still rough even with a chemical mechanic polishing step included. It is necessary to explore alternate growth approaches to achieve low defect density GaAs with a thin buffer layer.

In this paper, a thin strain-relaxed Ge buffer is grown by MBE to provide lattice matched substrates for GaAs growth. The Ge with low TDD is grown on Si through a two-step growth method, and the total thickness of Ge is 0.5µm. Single phase GaAs is grown on Ge buffers by using vicinal Si substrates and controlling the growth temperature and prelayers. Cross-sectional TEM (XTEM) and plain view TEM (PVTEM) are used to investigate the threading dislocation behavior and count the TDD in Ge and GaAs layers. TDD $<1\times10^8$ cm^{-2} is achieved for the GaAs on Si. The growth is monitored by in-situ reflective high-electron energy diffraction (RHEED), which shows a streaky 2×4 pattern for the GaAs growth on off-axis substrates. The 2×4 RHEED pattern indicates that single orientation GaAs is achieved, and XTEM further proved that the antiphase domain defects are negligible.

EXPERIMENTAL AND DISCUSSION

The growth of GaAs and Ge was carried out in two solid source Varian Gen II systems that were connected through an ultra high vacuum transfer tube. One system (system II) is used for the Si and Ge growth, and the other (system I) is for III-V epitaxy. Si substrates with 4° off (001) towards the <110> direction orientation were used in this experiment to get single phase GaAs on Ge.

A two-step growth without any anneal was first used for Ge growth. A thin Ge layer (50nm) was deposited at low substrate temperature (300°C), and then the substrate was heated to second stage temperature at about 600~800°C. The TDD of Ge was measured using plain view TEM (PVTEM) and the surface roughness was measured by atomic force microscopy (AFM). Figure 1a shows the cross sectional TEM image of sample grown by this two-step method. Under this growth condition, the strain relaxation process produced dense threading dislocations. Additional PVTEM shows the dislocation density is $1.4 \times 10^9 \text{cm}^{-2}$. The rms of the surface roughness for a 20×20µm^2 surface area is 1.4nm. The dislocation density slightly reduced when the 2nd stage growth temperature was increased, but the surface roughness increased rapidly.

To reduce the threading dislocation density, a high temperature anneal was used between the two growth steps. Figure 1b shows the cross sectional TEM of the Ge with one hour anneal at 1000°C. Most of the dislocations bent during the anneal step and few threading dislocations propagated throughout the Ge film, and thus a clean Ge layer without dislocations was achieved. The threading dislocation density was estimated to be $1 \times 10^8 \text{cm}^{-2}$ using plain-view TEM method. The rms of surface roughness for a 20×20µm^2 surface area was about 3.2nm.

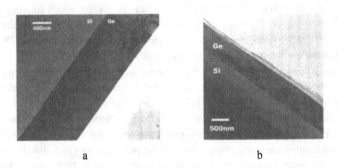

a   b

Figure1 XTEM of Ge on Si
a: Two-step growth without anneal;
b: Two-step growth with one hour anneal at 1000°C

After Ge film growth, the wafer was transferred to systemI under ultra high vacuum environment to grow GaAs. The substrate was heated to 500°C and then a monolayer of As_2 was deposited. 10 monolayers of GaAs was grown by 10 MEE steps before the codeposition of 1000Å GaAs at this temperature. The substrate was heated to 650°C to grow the thick structure. RHEED was used to monitor the growth. A streaky 2×4 RHEED pattern with 2× perpendicular to the atomic steps was observed, which indicated single phase GaAs with 4° off (001) towards (111)A was grown under this growth condition.

To investigate the growth quality of GaAs on Ge under the growth condition described above, two samples with different structure were grown. A simple $GaAs/Al_{0.1}Ga_{0.9}As$ structure was grown on GaAs substrate for the first sample. For the second sample, 20nm Ge was deposited on GaAs substrates first, and then the same $GaAs/Al_{0.1}Ga_{0.9}As$ structure was grown on Ge. (001) GaAs substrates with 4° offcut towards (111)A were used for both growth in order to obtain single phase GaAs on Ge. The second sample would provide a high quality Ge for the GaAs growth. The AFM results show both samples have very smooth surface with the rms of surface roughness at 0.2nm. Figure 2 shows the PL results of these samples. The main peak corresponds to $Al_{0.1}Ga_{0.9}As$ and the sub peak belongs to GaAs. The main peak intensity is same for two samples, but the GaAs peak of the sample with Ge below has higher intensity. This comparison indicates that very high quality GaAs has been grown on Ge under the growth condition.

GaAs was then grown on Ge-on-Si under the same growth condition. Figure 3a shows the XTEM image of GaAs on Ge, where the Ge was grown by a two-step method with 2nd stage temperature at 600°C and no anneal was used during the growth. The dislocations in GaAs were originated from the Ge substrates and no new dislocation was generated for the growth of GaAs on Ge. PVTEM shows the TDD in GaAs is 8× less than the Ge grown at the same condition.

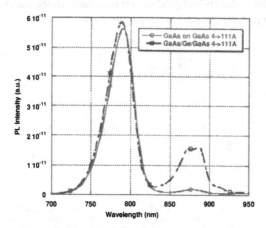

Figure 2 PL of GaAs on three types of substrates

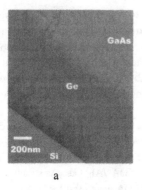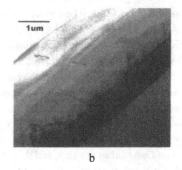

a

b

Figure 3 XTEM of GaAs on Ge-on-Si
a: No anneal during Ge growth;
b: 2hr anneal at 1000°C during Ge growth

Figure 3b shows the XTEM of GaAs on Ge where Ge was grown with 2nd stage temperature at 700°C with 2hr anneal at 1000°C. The dislocation in GaAs is $(5\sim10)\times10^7 cm^{-2}$ estimated by PVTEM. No antiphase defects are observed in figure3. The same structure was grown on GaAs substrate to compare the PL. Figure 4 shows the PL results. The PL of GaAs on Ge-on-Si is roughly 10× less than the PL of GaAs on GaAs substrates. The 10× reduction in PL intensity agrees with the result of the dislocation density estimation. Thus, to further improve the quality of GaAs on Si, the main challenge is still trying to produce a defect free Ge layer before the growth of GaAs. The two-step growth with additional anneal step can produce Ge layer with lower defect density and good surface roughness. Unfortunately, further increasing the anneal time didn't further reduce the defect density.

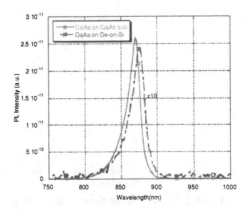

Figure 4 PL of GaAs on Ge-on-Si and on GaAs

CONCLUSION

The PL and AFM result indicates the growth quality of GaAs on Ge is as good as the GaAs grown on GaAs if low defect density exists in Ge layers. Thus, a high quality Ge film is needed to produce GaAs devices on Si with the same performance as the GaAs devices on GaAs substrates. Two-step growth and anneal has been used to grow Ge with TDD $<1 \times 10^8 cm^{-2}$ and the GaAs on this film yields a PL intensity about 10x less than the GaAs grown on GaAs. Further reduction of TDD in Ge is still needed to obtain GaAs on Si with the same quality as on GaAs substrates.

ACKNOWLEDGMENTS

This research is sponsored by Canon Incorporation. The authors would like thank Dr. Takao Yokohama in Canon for the instructive discussion. X. Yu also thanks the Winston and Fu-Mei Chen Stanford Graduate Fellowship, which has supported his graduate study at Stanford University.

REFERENCE

1. H. Kroemer. *J.Vac.Sci.Technol.B.* **5**, 1150 (1987)
2. Y. Takagi, H. Yonezu, Y. Hachiya and K. Pak. *Jpn. J. Appl. Phys.* **33**, 3368 (1994)
3. M.Shinohara. *Appl. Phys. Lett.* **52**, 543(1988)
4. N.A. El-Masry, J.C.L. Tarn and S.M.Bedair. *Appl. Phys. Lett.* **55** 1442(1989)
5. G.P. Watson, E.A.Fitzgerald,Y.Xie and D. Monroe. *J. Appl. Aphys.* **75**, 263(1994)
6. M.T.Currie, S.B.samavedam, T.A.Langdo, C.W.Leitz and E.A.Fitzgerald. *Appl. Phys. Lett.* **72**, 1718 (1998)

Mater. Res. Soc. Symp. Proc. Vol. 869 © 2005 Materials Research Society

Surface Acoustic Wave-Induced Electroluminescence Intensity Oscillation in Planar Light-Emitting Devices

Marco Cecchini[1], Vincenzo Piazza[1], Fabio Beltram[1], Martin Ward[2], Andrew Shields[2], Harvey Beere[3] and David Ritchie[3]
[1]Scuola Normale Superiore and NEST-INFM, I-56126 Pisa, Italy.
[2]Toshiba Research Europe Limited, Cambridge Research Laboratory, 260 Cambridge Science Park, Milton Road, Cambridge CB4 OWE, United Kingdom.
[3]Cavendish Laboratory, University of Cambridge, Cambridge CB3 0HE, United Kingdom.

ABSTRACT

Electroluminescence (EL) emission controlled by means of surface acoustic waves (SAWs) in planar light-emitting diodes (pLEDs) is demonstrated. Interdigital transducers (IDTs) for SAW generation were integrated onto pLEDs fabricated following a scheme compatible with SAW propagation [1]. EL in presence of SAW was studied by time-correlated photon-counting techniques. We found intensity oscillation at the SAW frequency (~1 GHz) demonstrating electron injection into the p-type region synchronous with the SAW wavefronts.

INTRODUCTION

Recently SAWs have attracted the interest of the semiconductor community in view of their interaction properties with two-dimensional-electron-gases (2DEGs) embedded in semiconductor heterostructures [2, 3]. Mechanical waves propagating across the surface of a piezoelectric substrate are associated with potential waves which couple with the free carriers confined in the quantum well. This interaction modifies the 2DEG equilibrium state and the wave speed and amplitude. By measuring the change in amplitude and velocity of the SAW it is possible to extract the frequency and wave-vector dependence of the 2DEG conductivity [2], while the momentum transferred by the SAW to the 2DEG leads to the excitation of dc current or voltage (the so-called acoustoelectric effect) [4, 5, 6].

The discovery of the acoustoelectric effect gave rise to the proposal of innovative devices. Among these Talyanskii et al. proposed the implementation of a novel current standard, demonstrating very precise acoustoelectric current quantization due to charge drag by SAWs through a quantum point contact [7, 8]. Control over the constriction width allowed very precise selection of the number of electrons packed in each SAW minimum down to the single-electron-transport regime. One of the most appealing applications proposed after the first report of the acoustoelectric quantized current was to incorporate single-electron SAW pumps in planar 2D electron/2D hole gas (n-p) junctions to fabricate single-photon sources [9].

In this paper we report on the acoustoelectric effect in pLEDs. Devices containing lateral *np* junctions and IDTs for SAW generation (SAWLEDs) were fabricated and measured at cryogenic temperatures. We analyzed the optical and transport properties induced by the acoustic perturbation by means of current-voltage (IV), light-voltage (LV) and EL time-resolved measurements. The SAWLED constitutes one of the main building blocks of the SAW-driven single-photon source [9] and, from a more fundamental physics point of view, allows the study of the acoustoelectric effect in planar systems where both electron and holes are present at equilibrium.

EXPERIMENTAL DETAILS

Devices were fabricated from a p-type modulation-doped $Al_{0.3}Ga_{0.7}As$/GaAs heterostructure grown by molecular-beam epitaxy, containing a two-dimensional-hole-gas (2DHG) within a 20-nm-wide GaAs quantum well embedded 70 nm below the surface. The measured hole density and mobility after illumination at 1.5 K were 2.03×10^{11} cm^{-2} and 35000 cm^2 / Vs, respectively.

The fabrication of the lateral junctions followed Ref. [1]. The heterostructure was processed into mesas with an annealed p-type Au/Zn/Au (5/50/150 nm) Ohmic contact. The fabrication of the n-type region of the junction consisted in the removal of the Be-doped layer from part of the mesa by means of wet etching (48 s in H_3PO_4:H_2O_2:H_2O = 3:1:50) and evaporation of a self aligned Ni/AuGe/Ni/Au (5/107/10/100 nm) n-type contact. After annealing, the penetrating n-type contact provided donors to the host semiconductor, creating an electron gas within the GaAs quantum well below the metal pad, adjacent to the 2DHG. The shape of n-contact was a thin stripe placed perpendicular to the SAW propagation direction, at a distance of 250 µm from the p-contact. The IDT was composed of 100 pairs of 200-µm-long Al fingers with 3-µm periodicity (~1 GHz resonance frequency on GaAs) and deposited at a distance of 800 µm from the mesa. The width of the n-type contact (2 µm) was of the same order of magnitude as the SAW wavelengths (3 µm) in order to limit SAW damping due to the massive metallization and SAW diffusion originating from nonuniform penetration of the Ohmic contact and inhomogeneities in the etched region. A schematic view of the device geometry is shown in Figure 1 (a).

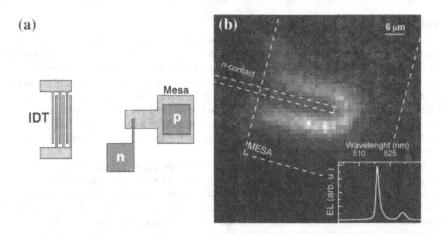

Figure 1. (a) Schematic view of the SAW-driven light-emitting device. (b) Spatially and spectrally resolved EL intensity profile of a lateral pn junction (T = 5 K) at a forward voltage of 1.7 V superimposed onto a schematization of the device. Inset: electro-luminescence spectrum at a forward bias of 1.8 V.

Several SAWLEDs were fabricated and measured obtaining qualitatively similar characteristics. All the data shown in the following refer to one representative device.

DISCUSSION

The transport and optical properties of the devices were preliminary tested without the presence of the SAW in order to verify the proper formation of the lateral junction. IVs were acquired at different temperatures from room temperature down to 5 K. The curves exhibited rectifying behavior with a threshold of ~1.5 V, as expected for GaAs np junctions. Emission properties were analyzed by EL measurements at low temperature (5 K). The spatial distribution of the emitted light was measured using an experimental setup based on a low-vibration cold-finger cryostat [1]. Figure 1 (b) shows the low temperature electroluminescence profile under 1.7 V forward bias superimposed onto a scheme of the device. Light emission originated within the mesa in proximity to the n-contact. Spectra were obtained by a cooled CCD after spectral filtering by a single-grating monochromator. The EL spectra were dominated by a strong main peak at 818.7 nm [full width half maximum (FWHM) of 1.8 nm] originating from radiative recombination within the QW [inset of Figure 1 (b)]. A secondary peak was also present at 831.2 nm with a FWHM of 3.6 nm due to carbon impurities unintentionally included in the heterostructure material during growth [10]. Light intensity measurements were performed by integration over the main EL peak, from 810 nm to 826 nm. As expected, the LV characteristics reflected the rectifying behavior of the IV curves, defining a detection threshold of ~1.65 V.

SAW excitation was verified by measuring the power reflected by the IDT as a function of the excitation frequency. The data acquired at 5 K showed a pronounced dip at 987.5 MHz (FWHM of 2.4 MHz), which is consistent with the expected SAW resonance frequency for a 3-μm IDT.

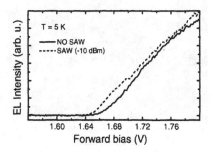

Figure 2. Light-voltage characteristic of the planar LED without SAW (solid line) and with SAW (dashed line). SAW power and frequency were -10 dBm and 987.5 MHz respectively.

Diode transport properties were significantly affected by the presence of the SAW [11]. We measured the variation (ΔI) of the current in presence of the SAW as a function of the power of the signal applied to the IDT (P_{RF}) at different forward voltages applied to the junction[1]. ΔI increased (i.e., more electrons injected into the 2DHG) for biases in the range of 1.6 to 2.0 V and for P_{RF} up to ~10 dBm. By increasing furthermore P_{RF}, ΔI was observed to decrease. This was probably due to sample heating originating from dissipation in the transducer and in the RF cable.

---

[1] We measured the current flowing out the n-type by low frequency lock-in techniques. The n-type contact was grounded.

The increased injection of electrons into the 2DHG due to the SAW is accompanied by a shift of the LV curve towards lower biases (see Figure 2). A maximum shift of 10 meV was obtained at $P_{RF} \approx$ -10 dBm. At higher power levels the increased electron extraction efficiency was counterbalanced by the spatial separation of electrons and holes trapped respectively in SAW minima and maxima which resulted in increased radiative lifetime [12] and thus in the suppression of the EL signal (inset of Figure 3). We observed a similar effect also in PL measurements (Figure 3 and inset). A region of the mesa was excited with a red-light laser source (653 nm) and PL was detected in presence of SAWs. The PL signal showed a monotonic decrease at increasing SAW field, demonstrating the efficiency of the SAW in turning off the PL even in presence of a 2DHG. Remarkably, the observed increase in the radiation intensity vs P_{RF}, for P_{RF} smaller than -5 dBm, was unique to the EL measurements. Emission intensity as a function of the frequency of the signal applied to the IDT intensified only for frequencies within the IDT passband, demonstrating the resonant nature of this effect. The emission-spectra features were changed by the SAW only in intensity and no spectral shift of the main peaks was observed.

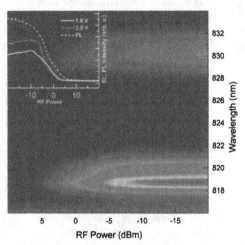

Figure 3. Photoluminescence spectra as a function of the RF power (P_{RF}) applied to the transducer (987.5 MHz). Inset: Photoluminescence and electroluminescence intensity as a function of P_{RF} at 987.5 MHz for two different forward voltages.

We finally analyzed the time evolution of the SAW-induced EL. The EL signal was spectrally filtered by a triple grating monochromator and detected by a single-photon APD module (Perkin Elmer SPCM-AQR-16). In order to obtain time-resolved EL traces we used time-correlated photon-counting techniques. The signal from the SPCM was directed toward the START input of a Becker & Hickl SPC-600 computer board, while the STOP input was driven by a signal synchronous to the SAW. Typical data traces showed 1-ns EL oscillations due to the SAW-induced modulation of the junction potential [13]. This effect is highlighted in the frequency domain of the signal. As shown in Figure 4, the Fourier transforms of the EL signals were indeed dominated by a single peak at the SAW frequency. The normalized oscillation amplitude was not observed to vary with bias voltage, as expected for an exponential increase of the EL close to the conduction threshold. The normalized amplitudes were quite low (about 1.5 % of the signal). Since the radiative recombination time in our system was measured by photoexcited-carrier lifetime measurements to be much less than 1 ns, we speculate that the main factor that limits the contrast of the oscillations originates from the detailed geometry of our devices. Ideally the SAW wavefronts would get to the thin n-type contact without any distortion and perfectly parallel to the contact itself. Nevertheless, real devices unavoidably present deviations from the ideal case. Mesa edges and defects along the SAW path introduced by the mesa etching process modified the acoustic wave-fronts. In addiction, annealed Ohmic contacts have a strong corrugation.

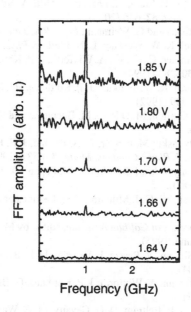

Figure 4. Fourier transforms of time-resolved electroluminescence traces in presence of SAWs (987.5MHz, -10 dBm), at T = 5 K and for different forward biases applied to the junction.

During the annealing process the metal diffusion made the junction edges irregular on the scale of the SAW wavelength. Electrons were thus extracted and recombined at different times depending on their position along the junction. The resulting oscillations were averaged out, while the EL mean-intensity increase still remained present. We believe that further optimization of the device geometry will lead to high contrast oscillations and is currently being investigated.

CONCLUSIONS

In conclusion, we studied the effect of SAWs on the emission properties of *np* lateral junctions. Electroluminescence was observed to increase in intensity in the presence of SAWs when the diode was biased near the conduction threshold. Time-resolved measurements of the electroluminescence in the presence of SAWs were carried out at several bias voltages, demonstrating modulation of the light intensity at the frequency of the SAW (~1 GHz). The amplitude of the oscillation was found to be ~1.5 % of the total light intensity, almost independent from the bias applied to the junction.

ACKNOWLEDGMENTS

This work was supported in part by the European Commission through the FET Project SAWPHOTON and through the FET IST SECOQC within FP6 and by MIUR within FISR "Nanodispositivi ottici a pochi fotoni".

REFERENCES

1. M. Cecchini, V. Piazza, F. Beltram, M. Lazzarino, M. B. Ward , A. J. Shields, H. E. Beere and D. A. Ritchie, *Appl. Phys. Lett.* **82**, 636 (2003).

2. A. Wixforth, J. P. Kotthaus and G. Weimann, *Phys. Rev. Lett.* **56**, 2104 (1986).

3. R. L. Willett, R. R. Ruel, K. W. West and L. N. Pfeiffer, *Phys. Rev. Lett.* **71**, 3846 (1993).

4. A. Esslinger, R. W. Winkler, C. Rocke, A. Wixforth, J. P. Kotthaus, H. Nickel, W. Schlapp and R. Lösch, *Surf. Sci.* **305**, 83 (1994).

5. A. Esslinger, A. Wixforth, R. W. Winkler, J. P. Kotthaus, H. Nickel, W. Schlapp and R Lösch, *Solid State Commun.* **84**, 939 (1992).

6. J. W. M. Campbell, F. Guillon, M. D'Iorio, M. Buchanan and R. J. Stoner, *Solid State Commun.* **84**, 735 (1992).

7. J. M. Shilton, V. I. Talyanskii, M. Pepper, D. A. Ritchie, J. E. F. Frost, C. J. B. Ford, C. G. Smith and G. A. C. Jones, *J. Phys.: Condens. Matter* **8**, L531 (1996).

8. J. Cunningham, V. I. Talyanskii, J. M. Shilton, M. Pepper, A. Kristensen and P. E. Lindelof, *Phys. Rev. B* **62**, 1564 (2000).

9. C. L. Foden, V. I. Talyanskii, G. J. Milburn, M. L. Leadbeater and M Pepper, *Phys. Rev. A* **62**, 011803(R) (2000).

10. B. Hamilton, in *Properties of Gallium Arsenide*, edited by M. R. Brozel and G. E. Stillman (INSPEC, London, England, 1996).

11. M. Cecchini, G. De Simoni, V. Piazza, F. Beltram, H. E. Beere and D. A. Ritchie, *Appl. Phys. Lett.* **85**, 3020 (2004).

12. C. Rocke, S. Zimmermann, A. Wixforth, J. P. Kotthaus, G. Böhm, and G. Weimann, *Phys. Rev. Lett.* **78**, 4099 (1997).

13. M. Cecchini, V. Piazza, F. Beltram, D. G. Gevaux, M. B. Ward, A. J. Shields, H. E. Beere and D. A. Ritchie, *cond-mat/0501136 v1* (2005).

Chemical and Biological
Sensing Systems

Mater. Res. Soc. Symp. Proc. Vol. 869 © 2005 Materials Research Society

A CMOS Medium Density DNA Microarray with Electronic Readout

Roland Thewes, Christian Paulus, Meinrad Schienle, Franz Hofmann, Alexander Frey,
Petra Schindler-Bauer, Melanie Atzesberger, Birgit Holzapfl, Thomas Haneder,
and Hans-Christian Hanke

Infineon Technologies AG, Corporate Research, D 81730 Munich, Germany

ABSTRACT

A CMOS chip-based approach is reviewed for fully electronic DNA detection. The
electrochemical sensor principle used, CMOS integration of the required transducer materials,
chip architecture and circuit design issues are discussed, respectively. Electrochemical and
biological results obtained on the basis of medium density microarray sensor CMOS chips with
16×8 sensor sites prove proper operation.

INTRODUCTION

In a wide-spread area of biotech and medical applications tools are required for the parallel
detection of specific DNA sequences in a given sample. Most prominent applications are genome
research and drug development, while applications in the area of medical diagnosis are under
development. Depending on the particular investigation, requirements range from simple
"presence or absence" tests to quantitative analyses with a relatively high dynamic range and
sensitivity.

So-called DNA microarrays [1-6], glass-, polymer-, or silicon-based slides with an active area
in the order of square millimeters to square centimeters, fulfill this request: There, different
species of single-stranded DNA receptor molecules (probe molecules) are immobilized at
predefined positions on the chip surface. To investigate a sample, the chip is flooded with the
sample containing the target molecules. Complementary sequences of probe and target molecules
hybridize, mismatching probe and target molecule do not bind. Finally, after a washing step,
double-stranded DNA is obtained at the match positions, and single-stranded DNA (i.e. the probe
molecules) remain at the mismatch sites. The information whether double- or single-stranded
DNA is found at a given test site reveals the composition of the sample, since the probes and
their positions on the chip are known. Consequently, sites with double-stranded DNA or the
amount of double-stranded DNA at the different sites, respectively, must be identified.

Commercially available state-of-the art DNA microarray chip systems use optical detection
techniques for that purpose [1-6]. By avoiding the relatively expensive and complicated optical
set-ups required there, electronic readout techniques in principle allow more robust and easier
operation. However, so far their status of development is lower.

Electronic approaches aiming for low density investigations (i.e. for applications with a low
number of sensors sites, e.g. of order 10), usually use a suitable bio-compatible chip substrate
material carrying the electrical transducer. The electrical terminals of each on-chip sensor are
directly connected to an off-chip reader system. Such "passive chips" represent a cost-effective
solution as long as only few sites per investigation are required. In case higher numbers of test
sites per chip are required this approach runs into an interconnect problem: The increasing

Table 1. Relationship between economic and technical aspects of passive and active electronic DNA microarrays.

	passive electronic DNA chips	active electronic DNA chips
density test sites per chip	low of order 10 (+/-)	medium ... high ≥ 100 (+/-)
costs per chip	low	increased processing costs: - CMOS processing costs - process to provide transducer elements must be compatible to CMOS process
cost per data point	approximately constant	decrease with - increasing number of test sites per chip - increasing number of required data points per investigation
electrical performance	medium	high
electronic signal integrity	- limited robustness - loss of signal integrity at high test site count per chip	- by far increased robustness - independent of number of test sites per chip

amount of interconnects lowers the available area per contact pad on-chip, so that contact reliability and yield decrease. Moreover, the limited available total chip area leads to a decreasing area per sensor which translates into decreasing signals.

To escape these drawbacks active on-chip circuitry is required, allowing to amplify and to process the weak sensor signals in the direct proximity of the sensors, and to communicate with the reader system on the basis of a low number of interconnects independent of the number of test sites per chip. Note, that although the costs per chip are higher in case of active chips, the costs per data point decrease, so that these represent the technically and economically better choice compared to their passive counterparts in case of medium and high density applications (Table 1).

In [7-11], the development of fully-electronic medium-density CMOS-based DNA microarrays is described. In the following chapters, we review the electrochemical sensor principle used, CMOS integration of the required transducer materials, chip architecture and circuit design issues, and consider electrochemical and biological results, respectively.

SENSOR PRINCIPLE

The electrochemical sensor principle used is schematically depicted in Fig. 1. It is based on an electrochemical redox-cycling technique [12-14]. A single sensor (Fig. 1, left) consists of interdigitated gold electrodes (generator and collector electrode). Probe molecules are spotted and immobilized (e.g. by means of Thiol coupling) on the surface of the gold electrodes.

The target molecules in the sample which is applied to the chip are tagged with an enzyme label (Alkaline Phosphatase). After the hybridization phase and a subsequent washing phase, a suitable chemical substrate (p-Aminophenylphosphate) is applied to the chip. The enzyme label, only available at the sites where hybridization occurred, cleaves the phosphate group and the electrochemically active p-Aminophenol is generated (Fig. 1, right).

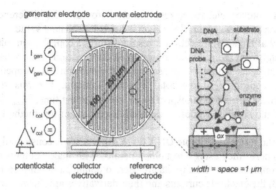

Fig. 1. Schematic plot showing the redox-cycling sensor principle and the sensor layout. Left: Single sensor composed of interdigitated gold working electrodes and potentiostat circuit with counter and reference electrodes. Right: Blow-up of a sensor cross-section showing two neighboring working electrodes after successful hybridization. For simplicity, probe and target molecules are shown on only one of the electrodes and are not to scale.

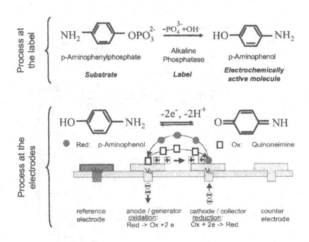

Fig. 2. Chemical processes at the label and at the electrodes.

Applying simultaneously an oxidation and a reduction potential to the sensor electrodes (V_{gen} and V_{col} in Fig. 5, e.g. +300 mV and −100 mV with respect to the reference potential), p-Aminophenol is oxidized to Quinoneimine at one electrode, and Quinoneimine is reduced to p-Aminophenol at the other one. The activity of these electrochemically redox-active compounds translates into an electron current at the gold electrodes (I_{gen} and I_{col}). The related chemical processes at the label and at the electrodes are summarized in Fig. 2.

Since not all particles oxidized at the generator reach the collector electrode, a regulated four electrode system is used (Fig. 1, left): a potentiostat, whose input and output are connected to a reference and to a counter electrode, respectively, provides the difference currents to the electrolyte, so that the potential of the electrolyte is held at a constant value.

The current flow at the sensor electrodes results from the contribution initially generated by the enzyme label, and from the redox-cycling related contribution at the sensor electrodes. Due to electrochemical artifacts within the phase, when the substrate is pumped over the chip, an offset current may contribute to the total detection current. For this reason, usually the derivatives of the sensor current with respect to the measurement time, $\partial I_{col}/\partial t$ and $\partial I_{gen}/\partial t$, are evaluated instead of the absolute values. A detailed discussion of this method is given elsewhere [12-14].

As an example, the result of a simple DNA oligo experiment is considered in the following. The measurement is performed using a test chip operated with analog outputs as discussed in [7, 10]. Relatively simple sensor site circuits are used to obtain an on-chip current gain of 100 (cf. Fig. 6). In Fig. 3, the electrode currents and their derivatives are shown for one position with matching sequences and for another position with mismatching sequences. As can be seen, the absolute values of collector and generator current and the absolute values of their derivatives are very similar. Match and mismatch clearly lead to different amounts of current within the evaluation time window. The derivatives, beyond that, cancel the effect of the offset currents and lead to data with opposite signs for the match / mismatch cases within the evaluation time window.

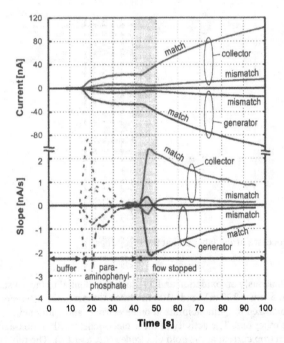

Fig. 3. Top: Measured sensor currents for one position with matching strands and for another position with mismatching sequences. Bottom: Derivatives of the sensor currents with respect to the measurement time. The evaluation time window is emphasized by a gray background.

CMOS INTEGRATION

The interaction of solid-state CMOS chips with the wet world of biology requires CMOS post-processing steps to provide the transducer materials such as gold or other noble metals in case of electrochemical principles [7-11, 15-19]. Processing of such materials directly within a CMOS production line is usually impossible due to contamination problems which have a significant impact on performance and yield of the CMOS devices.

In our case, the basic CMOS technology is a 5 V, 6" n-well process specifically optimized for analog applications with a minimum gate length of 0.5 μm and an oxide thickness of 15 nm. In order to provide the required gold electrodes, a Ti/Pt/Au stack (50 nm / 50 nm / 300-500 nm) is deposited and structured using a lift-off process. This post-CMOS process flow is schematically sketched in Fig. 4. SEM photos of the achieved results are given in Fig. 5.

However, also in the post processing case care must be taken, that the applied processing steps do not deteriorate the quality of the CMOS devices, in particular when sensitive analog circuitry is realized. As an example we consider sensor site test circuits designed to be operated with sensor currents from 1 pA to 100 nA. They consist of two regulation loops to control the bias voltages of both electrodes, whose currents are recorded and amplified by a factor of 100 using two cascode current mirrors in series. The simplified circuit diagram of one branch is shown in the inset of Fig. 6.

The circuits can be characterized using a test / calibration input [7-10]. Fig. 6 shows the relative error of the measured gain as a function of the input current (average value of all test sites from a 16×8 array chip). As can be seen, data taken from a wafer after gold deposition but without further processing steps reveal a strong deviation for input currents below 10 pA. Charge pumping characterizations [20] show, that this effect coincides with very high values for the gate oxide interface state density: values above 2×10^{11} cm^{-2} are obtained. Such values translate into an increased subthreshold slope of the transistors, worsened off-state characteristics, increased junction-to-substrate or junction-to-well leakage currents. Thus, they deteriorate the transfer characteristics for low currents (the most crucial circuit nodes are emphasized by asterisks in the inset of Fig. 6).

In order to shift this parameter back to reasonable values (e.g. of order 10^{10} cm^{-2}) forming gas annealing steps (N$_2$, H$_2$ at 400 °C / 350 °C, 30 min) are applied after gold processing. They significantly reduce the interface state density again and lead to reasonable transfer characteristics as depicted in Fig. 6 as well.

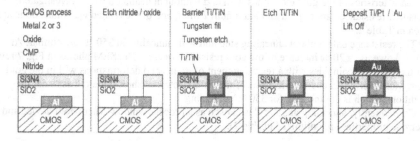

Fig. 4: Post-CMOS process flow to provide gold sensor electrodes.

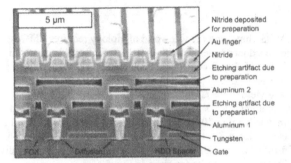
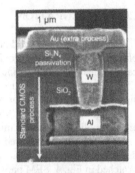

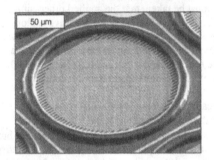
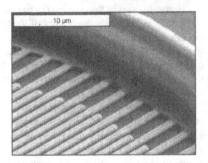

Fig. 5: SEM photos of the extended CMOS process. Top left: Tilted cross section with Au sensor electrodes and CMOS elements after the complete process run. Note that the nitride layer on top of the sensor electrodes is only used for preparation purposes. Top right: Blow-up showing Au electrode, last Al-layer from the CMOS process, and Al-to-Au via. Bottom left: Top view of a sensor with interdigitated gold electrodes (embedded within a Polybenzoxazole compartment ring). Bottom right: Top view blow-up showing part of a sensor with interdigitated gold electrodes and compartment ring.

However, this is not the entire story: In addition to the CMOS process front-end parameters, the characteristics of the gold electrodes with and without annealing must be investigated. Measured resistance data of gold and aluminum 2 lines, and of the related via connections are given in Table 2.

The resistance data without annealing step and with annealing at 350 °C are similar. At 400 °C however, a 20 % increase of the gold resistance occurs. The SEM photos in Fig. 7 reveal that this increase coincides with a rearrangement of grains and deformations within the gold layer. Moreover, yield characterizations of the gold electrodes show a yield loss under these conditions which is not the case for 350 °C annealing.

Consequently, annealing at 350 °C is chosen as a process window where both device and electrode properties are optimized (Table 2).

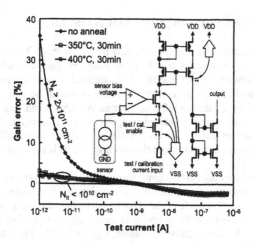

Fig. 6: Error of current gain of the circuit shown in the inset as a function of the input test current normalized to the gain at test current = 1 nA for different annealing options after gold processing. N_{it}: measured interface state densities.

Table 2: Measured resistance data of gold and aluminum 2 lines, of the related via connections, and of the gate oxide interface state density for different annealing options after gold processing.

	square resistance Au lines [mΩ/square]	resistance via holes (Al to Au) [mΩ]	square resistance Al 2 lines [mΩ/square]	interface state density [1/cm²]
CMOS only (i.e. without Au process)	-	-	-	~ 10^{10}
CMOS + Au process, no anneal	48	370	79	~ 2×10^{11}
CMOS + Au process, N₂/H₂ anneal with 350°C, 30 min	51	360	76	< 10^{10}
CMOS + Au process, N₂/H₂ anneal with 400°C, 30 min	61	340	74	< 2×10^{9}

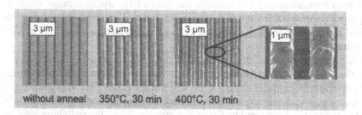

Fig. 7: SEM photos showing the Au sensor electrodes without and with annealing steps performed at different temperatures after Au processing.

CHIP ARCHITECTURE AND CIRCUIT DESIGN ISSUES

Fig. 8 depicts the architecture of a user-friendly prototype array. The related chip photo is shown in Fig. 9. The chip consists of 16×8 sensor positions. They are specified to be operated with sensor currents between 10^{-12} A and 10^{-7} A. Electrode bias voltages are provided to each sensor site by two digital-to-analog converters in the periphery of the chip. All references are generated or derived from a bandgap reference circuit. A global potentiostat is used. The sensor sites can be electrically tested and calibrated by application of calibration currents, which are generated by the calibration engine. The cost of deriving high accuracy calibration currents over five decades from a reference current from the bandgap circuit is the large area consumption of this block [21, 22].

The chip is connected to the reader device by six pins. Two of them are used for power supply, the remaining four pins for serial data communication. The related serial interface logic circuitry controls the operation mode of the chip, the x- and y-decoders to select the sensor sites, and a number of registers to store the control data for the digital-to-analog converters, the calibration engine, ... A more detailed discussion is given in [11]. The most important data are summarized in Table 3.

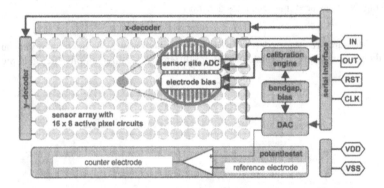

Fig. 8: Architecture and organization of the prototype array with 16×8 sensor positions.

Sensor site operation at very low currents and over a large dynamic range suggests to leave the analog signal representation and to switch to the digital domain as early as possible to ensure robust signal transmission through the array, i.e. high signal integrity, and high crosstalk and disturb immunity.

This is achieved by in-sensor site analog-to-digital conversion using the concept depicted in Fig. 10 for the collector branch: The voltage of the sensor electrode is controlled by a regulation loop via an operational amplifier and a source follower transistor. For analog-to-digital conversion, a current-to-frequency converting sawtooth generator concept is used, where an integrating capacitor C_{int} is charged by the sensor current. When the switching level of the comparator is reached, a reset pulse is generated which passes through the delay stage and the capacitor is discharged by transistor M_{res} again. The delay stage is required as a pulse shaping

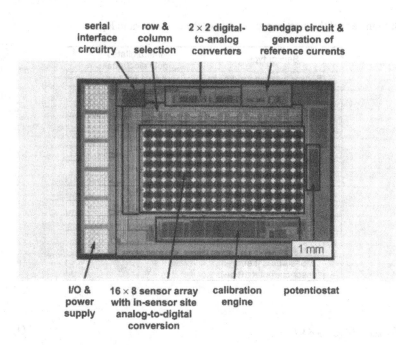

serial interface circuitry   row & column selection   2 × 2 digital-to-analog converters   bandgap circuit & generation of reference currents

1 mm

I/O & power supply   16 × 8 sensor array with in-sensor site analog-to-digital conversion   calibration engine   potentiostat

Fig. 9: Chip photo of the prototype array with 16 × 8 sensor positions.

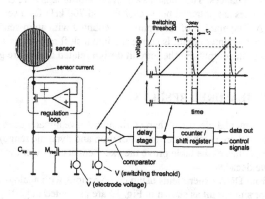

Fig. 10: Circuit principle used for in-sensor site analog-to-digital conversion of the sensor signal based on current-to-frequency conversion.

unit to ensure a sufficient length of the reset pulse applied to M_{res} so that complete discharging of the integrating capacitor is guaranteed. The number of reset pulses is counted with a digital counter. The measured frequency approximately amounts to

Table 3: Summary of the most important data of the chip shown in Fig. 9.

Sensor type	Interdigitated gold electrodes
Sensor electrode width	1 µm
Sensor electrode spacing	1 µm
Sensor diameter	150 µm
Sensor pitch	250 µm
Sensor count	16 × 8
Total chip size	6.4 mm × 4.5 mm
CMOS process	6", n-well, L_{min} = 0.5 µm, t_{ox} = 15 nm
Supply voltage	5 V
Power consumption	~ 200 mW
Current resolution	10^{-12} A ... 10^{-7} A
Measurement error (sensor current > 10 pA)	σ < 1 %
Measurement error (sensor current < 10 pA)	σ < 3 %

$$f = I_{electrode} / (V_{ref,comp} \times C) \tag{1}$$

with $I_{electrode}$ being the electrode current, $V_{ref,comp}$ the comparator switching level, and $C = C_{int}$ + parasitic capacitances. The chosen switching amplitude is 1 V, C is approximately 140 fF, so that frequencies are obtained between 7 Hz and 700 kHz for a sensor current range from 10^{-12} A to 10^{-7} A. The number of reset pulses is counted with an in-sensor-site 22-stage counter. For readout, the counter circuit is converted into a shift register by a control signal and the data are provided to the output. Further circuit design related details are given in [9].

EVALUATION OF MEASURED RESULTS

Results from experimental electrical characterizations have already been provided in Table 3. All specifications given there (such as current resolution and dynamic range) are fulfilled or exceeded. The last two lines of the table prove the excellent accuracy of the chip over the large dynamic range of five decades.

Measured data from DNA experiments using chip versions with analog sensor sites and on the basis of the sensor site circuit as shown in Fig. 10 are presented in [7-10] (cf. also Fig. 3). They prove proper functionality of the chip under biological operation.

A detailed electrochemical evaluation of the chip shown in Fig. 9 is presented in [11]. There, p-Aminophenylphosphate is applied to the chip in different concentrations representing typical values as obtained in DNA experiments. An approximately linear response is obtained over nearly four decades. The achieved sensitivity and dynamic range is excellent for most typical applications.

SUMMARY

An electrochemical redox-cycling based DNA detection principle using interdigitated gold electrodes is extended for application on CMOS chips, which are mandatory for medium and high density investigations. Integration and post processing issues of the transducer material gold have been discussed in detail. Architecture and organization of a user-friendly prototype array with 16×8 sensor positions as well as circuit design details have been considered. Experimental electrochemical and biological results prove proper operation of the considered approach.

REFERENCES

[1] E. M. Southern, Anal Biochem., 62(1), p. 317, 1974

[2] http://www.nature.com/ng/chips_interstitial.html

[3] "DNA microarrays: a practical approach", M. Schena ed., Oxford University Press Inc., Oxford, UK, 2000

[4] "Microarray Biochip Technology", M. Schena ed., Eaton Publishing, Natick, MA 01760, 2000

[5] D. Meldrum, Genome Research, 10, p. 1288, 2000

[6] F. Bier et al., in 'Frontiers in Biosensorics I', F. Scheller et al. ed., Birkhäuser Verlag Basel/Switzerland, 1997.

[7] R. Thewes et al., Tech. Dig. ISSCC, p. 350, 2002

[8] F. Hofmann et al., Tech. Dig. IEDM, p. 488, 2002

[9] M. Schienle et al., IEEE Journal of Solid-State Circuits, p. 2438, 2004

[10] "CMOS-based DNA sensor arrays", R. Thewes et al., in 'Advanced Micro and Nano Systems – Enabling Technology for MEMS and Nanodevices', H. Baltes et al., Wiley-VCH, 2004.

[11] A. Frey et al, accepted for publication in Proc. ISCAS, May 2005

[12] R. Hintsche et al., in 'Frontiers in Biosensorics I', F. Scheller et al. ed., Birkhäuser Verlag Basel/Switzerland, 1997

[13] A. Bard et al., Anal. Chem., 58, p. 2321, 1986

[14] M. Paeschke et al., Electroanaysis, 8, No. 10, p. 891, 1996

[15] K. Dill et al., Anal. Chim. Acta, 444, p. 69, 2001

[16] K. Dill et al., J. Biochem. Biophys. Methods, 59, p. 181, 2004

[17] C. Paulus et al., Proc. IEEE Sensors, p. 474, 2003

[18] T. Sosnowski et al, Proc. Natl. acad. Sci. USA, Vol. 94, p. 1119, 1997.

[19] M. Heller, IEEE Engineering in Medicine and Biology Magazine, p. 100, 1996

[20] G. Groeseneken et al., IEEE Trans. Electron Devices, p. 42, 1984

[21] M. Pelgrom et al., IEEE Journal of Solid-State Circuits, p. 1433, 1989

[22] K. Laksmikumar et al., IEEE Journal of Solid-State Circuits, p. 657, 1985

Mater. Res. Soc. Symp. Proc. Vol. 869 © 2005 Materials Research Society   D3.2

MACROPOROUS SILICON SENSOR ARRAYS FOR CHEMICAL AND BIOLOGICAL DETECTION

Jeffrey Clarkson, Vimalan Rajalingam, and Karl D. Hirschman
Departments of Microelectronic Engineering and Materials Science & Engineering,
Rochester Institute of Technology, Rochester, NY 14623

Huimin Ouyang, Wei Sun and Philippe M. Fauchet
Departments of Biomedical Engineering and Electrical & Computer Engineering, University of
Rochester, Rochester, NY 14642

ABSTRACT

A new class of silicon-based chemical and biological sensors that offer an electrical response to a variety of substances is described. The devices utilize silicon flow-through sensing membranes with deep trench structures formed to depths up to 100µm, fabricated by electrochemical etching which transforms the silicon into macro-porous silicon (MPS). The sensors have demonstrated the ability to detect the presence of certain chemical and biological materials. Although the principle of operation of the devices is fairly complex, the transduction mechanisms can be compared to chemiresistors and chemically sensitive field-effect transistors (chemFETs). The electrical responses that have shown the most sensitivity are AC conductance and capacitance. Previous work has demonstrated that upon exposure to organic solvents (i.e. ethanol, acetone, benzene) the devices exhibit a characteristic impedance signature. The devices have also shown the ability to detect the hybridization of complementary DNA. The incorporation of other materials that have demonstrated sensitivity to low ambient levels of contaminants is also under investigation. The sensors have been designed and fabricated in linear array configurations; a microfluidic transport chip/package co-design is currently in progress.

INTRODUCTION

There have been several types of gas and chemical sensors developed, each with their own specific advantages and disadvantages in performance with respect to sensitivity, selectivity, power consumption, size, cost, and applications (Ref. 1 provides a thorough review). Chemi-resistors and ChemFETS are among the available solid-state devices that have advantages over other classes (e.g. optical) in size and portability, but continue to have issues related to selectivity and sensitivity [1]. With the increased emphasis on homeland security, the ability to detect biological materials has become increasingly important. Although electronically integrated microarrays for DNA analysis have been developed, commercial technologies rely primarily on off-chip optical detection / readout platforms [2]. Integrated optical detection schemes have been implemented [3], although the system complexity increases significantly. Electrical readout schemes are currently being developed [4], which offer significant advantages in system integration.

Silicon has been demonstrated to be promising as a biocompatible material [5], and has an unmatched capability to provide a platform for integrated electronics and mechanical structures. There have also been numerous reports of porous silicon applications in chemical and biological sensing [6-9]. Of the various forms that porous silicon can take, macro-porous silicon (MPS) has pores with diameters between 1-2µm that can extend deep into the substrate. This material has demonstrated sensitivity to certain molecular processes [10,11], and can facilitate a flow-through membrane structure for liquid-phase and gas-phase detection. This paper will report on the use of MPS in a new device architecture, which utilizes a flow-through sensing membrane for microfluidic system integration.

EXPERIMENTAL

The sensors were developed by forming a surface layer of MPS, which intersects a backside KOH-etched cavity forming a flow-through membrane. Aluminum contact electrodes were placed on the membrane structure, however in a non-flow-through region which does not come into contact with the analyte under test. Prior to contact formation, the membrane was thermally oxidized, lining the pores with SiO_2 which ensures electrical isolation between the analyte and contact electrodes, and provides a hydrophilic internal surface. This section will review the basic fabrication process of the sensing membrane, and the development of the flow-through sensing structure.

Flow-through Macroporous Silicon Membrane

MPS sensor devices were fabricated on 4inch (100) oriented boron-doped p-type silicon wafers within a resistivity range $\rho \sim 20\text{-}25\Omega$cm. The MPS layer was formed by electrochemical dissolution under galvanostatic conditions using a current density of $4mA/cm^2$. The electrolyte used was 4% wt. hydrofluoric acid (49 % wt.) in N,N-Dimethylformamide (DMF). The use of a mild oxidizer such as DMF with low doped silicon results in very straight pore walls and pore diameters between 1-2μm [12].

The membranes were etched to approximately 70μm deep in select regions using silicon nitride as an etch mask. Another nitride mask was defined on the backside of the wafer (aligned to the front) and a KOH etch was used to form a deep cavity, providing fluidic delivery to the free-standing membrane. Finally, the structure was thermally oxidized, providing an SiO_2 layer approximately 1000Å thick within the macropores.

Contact Electrodes & Fluidic Channel Arrays

Contact cuts through the passivation oxide were patterned and etched, and aluminum electrodes were deposited and patterned. Note that this sensor configuration allows the sensor contacts, and potentially integrated microelectronics, to be isolated from the backside fluidics, as shown in figure 1. An overcoat layer is deposited and openings are patterned and etched to provide access to the electrodes and flow-through regions. The KOH V-groove can extend over several devices, and fluidic channel arrays were fabricated as shown in figure 2. Although there can be some parasitic influence, the isolation is adequate for each opposing electrode pair to sense independently from a neighboring device.

Electrical Characterization

The sensing membrane can be modeled as a distributed RC network. There are conductive pathways through the interconnected silicon regions, along with capacitive coupling across insulating regions; the effective medium composed of SiO_2, void space and depleted silicon. Capacitance is typically used as the

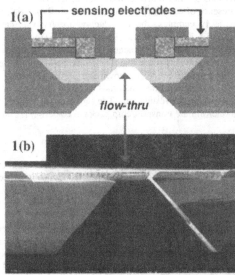

Figure 1. Sensor schematic and SEM cross-sections. The schematic (a) shows the flow-through sensing membrane, with contact electrodes isolated from the flow-through region. The SEM image (b) shows the actual scale of the sensing membrane and the KOH-etched cavity beneath. Note that in this sample the KOH-etched cavity overextends significantly into the MPS layer.

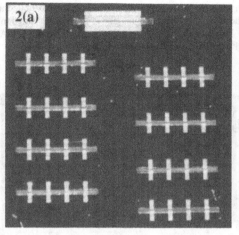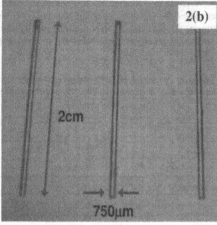

Figure 2. Linear-array configuration of sensors. The frontside view (a) shows four opposing electrode pairs per channel, with a full-area contact device for measurement comparison. Each individual sensor has a baseline capacitance ~ 2nF, in comparison to the full-area device capacitance of ~ 40nF. The backside view (b) shows the fluidic channels for analyte delivery. The channel length is ~ 2cm, with a channel width of ~ 750μm. Note that the actual flow-through region of the membrane is significantly narrower (~ 100μm).

measured electrical response to various chemical and biological materials; an electrical model for the sensing mechanism has been developed [13,14]. Capacitance measurements were performed in individual PSi sensing elements by connecting them through the lateral aluminum pads to an HP-4275A multi-frequency LCR meter. Real time data acquisition and storage was done by a computer controlled interface with LabView™. The measurements were performed at 10 kHz with a 60mV pp AC signal at 0V DC bias.

RESULTS

A number of substances were tested with the developed sensors; both in liquid-phase and vapor-phase. The sensors are quite sensitive to changes in the ambient conditions such as relative humidity, temperature and light; these parameters were monitored during testing to ensure the device response was exclusively from the analyte. The initial study involved liquid-phase testing of solvents with various dielectric properties (i.e. relative permittivity, dipole moment). Repeatability (repeated trials of a single solvent / single device combination) and reproducibility (single device testing multiple solvents) was also examined. Vapor-phase testing of select solvents was then investigated at both strong and dilute ambient concentrations. Selective binding of proteins (biotin & streptavidin) was also explored within buffer solution, due to the possibility of a detectable signal upon protein immobilization within the macropore structure.

Liquid-Phase Solvent Detection

Table 1 shows the electronic properties of the solvents tested along with the vapor pressure which determines the rate of evaporation, and thus the time duration of the response signal. The device chips were placed into the test fixture of the LCR meter, channel side up, and a baseline capacitance measurement was taken. Approximately 30μL of solvent was introduced in the center of the channel, filling the channel and sensing membrane. Figure 3 shows the real-time measurement, where the signal is represented by a change in capacitance (ΔC) compared to the baseline value. Each solvent exhibited a characteristic signature, with the exception of toluene which lacked a signal above the natural variation (noise) in the sensor response due to its low dipole moment and relative permittivity [14]. A demonstration of the repeatability and reproducibility of the sensors is shown in figure 4.

Table 1. Properties of Solvents Investigated

Material	Dielectric constant	Dipole Moment (D)	Vapor Pressure (T)
Acetone	21	2.88	233
Isopropanol	18.3	1.66	59
Ethanol	24	1.69	43
Methanol	33	1.7	128
Toluene	2.4	0.36	29

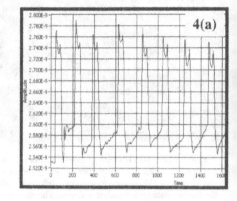

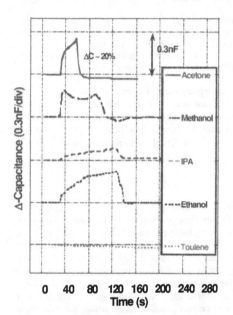

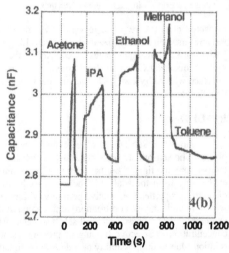

Figure 3. Characteristic signatures of the solvents tested. Each measurement was taken from a sensor specifically dedicated to that specific solvent. Acetone exhibited the strongest signal, with a change in baseline capacitance of ~ 15%. Each signature is unique to the specific solvent, with distinctive features such as the coincidental "M" characteristic of the methanol waveform.

Figure 4. Repeatability and reproducibility of the liquid-phase solvent response. (a) Methanol characteristics taken repeatedly on the same sensor, exhibiting consistent peak measurement and return levels. (b) Five solvent characteristics taken sequentially on the same sensor. Although the return response level shifts upwards, the response signatures are virtually the same as that shown in figure 3 taken from sensors dedicated to a single solvent.

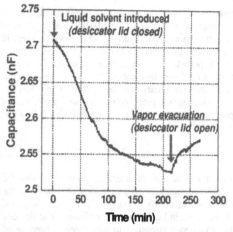

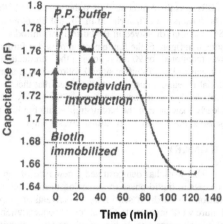

Figure 5. Sensor response to 100ppm ethanol vapor. The characteristic shows a decrease in capacitance, unlike the response to a saturated vapor ambient. As the liquid ethanol evaporates the signal decreases at a near constant rate until ~ 100min, after which the rate of signal change is reduced. The signal shows an immediate increase upon vapor evacuation once the desiccator lid is removed at 220min.

Figure 6. Sensor response to protein binding. Aminosilane (2% in acetone) was used as the linker to the passivating oxide layer. Biotin was immobilized (1mg per mL PBS, 10min) followed by a PP-buffer rinse. After a stabilization period, streptavidin (1mg per mL PP-buffer) was introduced. Note that the sensor signal rise above the measurement of ~ 1.76nF prior to streptavidin introduction was due to the desiccator lid positioned in place; the dip at ~ 10min is due to a brief removal, and the drop at 20min follows lid removal. The lid remained in place following introduction of the streptavidin at ~ 32min.

Vapor-Phase Solvent Detection

The response of the sensors to solvents in vapor-phase was then investigated. Initial tests were performed with a saturated vapor ambient; liquid solvent beneath the sensor in a 1.5L desiccator. Although there were subtle differences in the response characteristics, each rose gradually and continued to increase until saturation after several hours. As the ambient concentration was reduced, differences in response became more apparent. Figure 5 shows the sensor response to ethanol vapor at an ambient concentration of approximately 100ppm (0.15μL evaporated in 1.5L contained air ambient). The sensor used for this test was the same sensor dedicated for the liquid-phase ethanol characteristics. As the ethanol evaporated the capacitance demonstrated a decrease in response to the low vapor concentration, with the signal change slowing markedly after 100min. Upon removal of the desiccator lid there was a graduate increase in the response, with eventual return to near 2.7nF (not shown). This procedure was repeated using the same solvent ambient and sensor combination, producing the same response characteristic within noise limitations. At this point the dependence of the sensor behavior on the solvent ambient concentration it is not well understood; this is under further investigation.

Selective-Binding Biological Materials

In previous work [10,11,13] the sensor has shown the capability of detecting DNA hybridization. In this study the selective binding of streptavidin and biotin was investigated. The first step in the procedure was to perform a surface pretreatment of the SiO_2 passivation layer with aminosilane, which results in the attachment of amine groups (NH_3) to the oxide surface. Biotin in a phosphate buffered saline solution

(PBS) was then introduced in to the sensor, and 30min was allowed for attachment to the amine groups. The sensors were then rinsed in PP-buffer solution (phosphate buffer containing 0.01% NaN$_3$, pH 6.0). Streptavidin (dissolved in PP-buffer) was introduced after 30min of stabilization. The response signal is shown in figure 6; the figure caption provides further details on the sequence of events. The sensor signal was monitored for 90min following the introduction of streptavidin. The signal showed an initial increase (similar to the PP-buffer signal) followed by a significant decrease, leveling off at ~ 6% lower then the initial baseline value. It must be noted that these results are preliminary and have yet to be reproduced; most likely due to inconsistencies in the quality of the oxide surface pretreatment. However, the results clearly indicate that the biotin/streptavidin binding process induces a change that can be detected by the sensors; this is also under further investigation.

SUMMARY

This work has demonstrated a new type of sensor that can be used to detect the presence of specific solvents in liquid phase in an analog fashion due to response signature characteristics. In addition, the sensors also exhibit sensitivity to solvents in vapor phase under low (100ppm) ambient conditions. Future work involves testing at lower concentration levels, both in liquid and vapor phases. The devices also look promising as biological sensors, exhibiting a response attributed to the selective binding of streptavidin to immobilized biotin that was markedly different than the sensor response to the PP-buffer solution alone. Additional experiments are currently in progress to verify the results observed. The devices offer a flow-through membrane for fluidic transport, and effective isolation of the sensing electrodes from the materials under investigation. An electronic/fluidic package is currently in development. Further size reduction and the future integration with microelectronics and microfluidics will mark significant steps toward the realization of a lab-on-a-chip microsystem.

REFERENCES

[1] D.M. Wilson, S. Hoyt, J. Janata, K. Booksh, and L. Obando, *IEEE Sensors Journal* 1 256-274 (2001).

[2] P. Swanson, R. Gelbart, E. Atlas, L. Yang, T. Grogan, W.F. Butler, D.E. Ackley and E. Sheldon, *Sens. Actuators B* **64** 22-30 (2000).

[3] T. Vo-Dinh, G. Griffin, D.L. Stokes and A. Wintenberg, *Sens. Actuators B* **90** 104-111 (2003).

[4] L.F. Fuller, R. Vega, R. Manley, V.C. Hwang, D. Jaeger, A. Pham, N. Wescott and M. Connolly, *IEEE UGIM Proc.* **15** 200-202 (2003).

[5] L. Canham and R. Aston, *Physics World* **14**, 27-32 (2001).

[6] S.E. Létant, S. Content, T. Tan, F. Zenhausern and M.J. Sailor, *Sens. Actuators B* **69**, 193-198 (2000).

[7] S. Belhousse, H. Cheraga, N. Gabouze, and R. Outamzabet, *Sens. Actuators B* **100** (2004) 250–255 (2004).

[8] R. Angelucci, A. Poggi, L. Dori, G.C. Cardinali, A. Parisini, A. Tagliani, M. Mariasaldi, and F. Cavani, *Sens. Actuators A* **74** 95–99 (1999).

[9] H. Ouyang, L.A. DeLouise, M. Christophersen, B.L. Miller and P.M. Fauchet, *Proc. SPIE* **5511**, 71-80 (2004).

[10] M. Archer , P.M Fauchet, *Phys. Stat. Sol. (a)*, **198**, 503-507 (2003).

[11] M. Archer, M. Christophersen and P.M Fauchet, *Mat. Res. Soc. Symp. Proc.* **737** 549-554 (2003).

[12] H. Foell, M. Christophersen, J. Carstensen and G. Hasse, *Mat. Sci. and Eng.* **R39** 93-139 (2002).

[13] M. Archer, M. Christophersen and P.M. Fauchet, *Biomedical Microdevices* **6** 203-211 (2004).

[14] M. Archer, M. Christophersen and P.M. Fauchet, *Sens. Actuators B* **106** 347–357 (2005).

Mater. Res. Soc. Symp. Proc. Vol. 869 © 2005 Materials Research Society

Monolithic Liquid Chemical Sensing Systems

Steven M. Martin[1,2], Timothy D. Strong[1,3], and Richard B. Brown[1,4]
[1]University of Michigan, Ann Arbor, MI 48105, USA
[2]Sonetics Ultrasound, Ann Arbor, MI 48103, USA
[3]Sensicore, Ann Arbor, MI 48108, USA
[4]University of Utah, Salt Lake City, UT 84112, USA

ABSTRACT

The miniaturization of electrochemical transducers is crucial for the development of implantable biosensors, wearable microdetectors, and remote sensing networks due to the small dimensions required in these applications. As sensors scale, the analytical response degrades and integrated instrumentation is required to maintain acceptable detection limits. This work details the development of CMOS-integrated liquid chemical sensors. The sensors were cost-effectively post-processed on top of foundry-fabricated CMOS electronics using thin-film techniques. CMOS-integrated voltammetric sensors demonstrated a 25x improvement in detection limit/electrode area versus passive sensors. CMOS-integrated, ion-selective electrodes demonstrated a 50x improvement in lifetime and a 200x improvement in response time versus passive sensors. With their improved performance, these smart sensors can be used in a wide range of applications and can additionally serve as enabling technologies for more complex, chip-scale systems.

INTRODUCTION

By integrating electrochemical sensors and signal conditioning circuitry onto a monolithic substrate, the sensor's minimum detectable signal can approach theoretical noise limits since the integrated amplifiers and impedance transformation circuits reduce the ambient noise coupled into the system. Additionally, these monolithic microsystems can extend the sensor's lifetime, improve linearity, and enable dimensional scaling. Thus, integration facilitates the application of these microsystems to environmental monitors, portable handheld instruments, and implantable devices. The fabrication costs of these systems, however, have remained a concern as many integrated systems require custom CMOS fabrication technologies [1,2]. In this work, we discuss the development of a cost-effective fabrication technology where the electrochemical sensors are post-processed on top of foundry-fabricated CMOS substrates. Two chemical microinstruments developed using these techniques are described and results from their characterization are given. These devices are also compared with passive transducers.

Two types of electrochemical sensors are discussed in this work. The first group of sensors is amperometric or voltammetric sensors. When a potential is placed on an amperometric sensor, a faradaic reaction will occur if the voltage exceeds the standard potential of a solution constituent. This creates a current that is proportional to the concentration of that analyte. By employing a continuously varying applied potential, an amperometric sensor can be used in a voltammetric configuration. A full description of amperometric and voltammetric sensors is beyond the scope of this work, but the interested reader is directed to [3] for more information. The voltammetric sensors employed here consist of three electrodes. The faradaic reaction of interest occurs at the working electrode (WE). The reference electrode (RE) tracks the solution potential and is used for closed-loop control of the applied potential. Finally, the auxiliary electrode (AE) serves to close the current path through the solution and assists in the closed-loop control of the solution and electrode potentials.

The second class of transducers used in this work is potentiometric sensors. Potentiometric sensors, or ion-selective electrodes (ISEs) accumulate a charge on their surface which is proportional to the activity (approximately the concentration) of an analyte in solution. Ideally, this is a selective process where the charge accumulation is due only to an analyte for which the electrode has been selectively doped. When measured versus the solution potential, this charge creates a voltage. Typical solid-state ISEs have extremely large output impedances [4] and must be operated under equilibrium conditions, i.e. zero current draw. Further details on ISEs can be found in [5].

FABRICATION OF CMOS-INTEGRATED SENSORS

A thin-film technique for the fabrication of ISEs and amperometric sensors without CMOS electronics was described in [6]. That process served as the basis for the development of a post-CMOS fabrication technique for integrating liquid chemical sensors on top of foundry-fabricated CMOS electronics [7,8]. A cross-section of a completed CMOS-integrated microinstrument is shown in Figure 1. The remainder of this section details the steps involved in the manufacturing process.

CMOS Substrate

The CMOS electronics were fabricated in AMI Semiconductor's *(Pocatello, ID)* C5, 0.5μm, mixed-signal, CMOS process using that process's standard design rules. Alignment features for a 5 mask post-CMOS processing procedure were laid out using the top metal layer. In addition to the nominally-sized bonding pads, 960μm^2 contacts were placed near the center of the die to serve as connections to the sensor sites. Through the MOSIS service [9], the completed electronics were delivered as individual die measuring 71.5mm^2. This rather large device

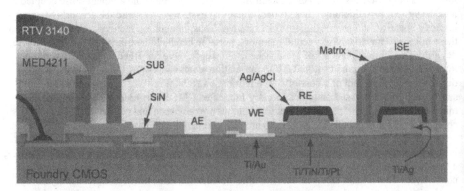

Figure 1. Cross section of CMOS-integrated liquid chemical sensors.

allowed for handling with tweezers without damaging the electronic structures in the center of the die. This extra handling area is only necessary for die-level processing and is not required for wafer-level processing. The completed MOSIS die contained a fully functional set of sensor interface circuits as described in [8].

CMOS/Sensor Interface

The top-level metal of a standard CMOS process (typically aluminum or an aluminum mixture) is incompatible with the metallization required for liquid chemical sensors such as platinum and gold. Under heating, the aluminum and other metallization layers hillock which renders the devices useless. A diffusion barrier is required between the layers. We first attempted to use TiW as a diffusion barrier, but large, additive stresses between the TiW and Pt caused adhesion problems during processing. TiN was used as a substitute material and showed less severe additive stress [7].

First, photoresist was patterned for the metal lift-off procedure. Then a layer of 40/80nm Ti/TiN was reactively sputtered onto the surface of the wafer followed by a 20/100nm Ti/Pt evaporation. This platinum layer serves as both a sensor interconnect layer and also as the electrode material for the auxiliary electrode. Platinum was chosen as a suitable material due to its wide electrochemical potential window in aqueous solutions [10]. The Ti adhesion layers were found to reduce defects induced in subsequent heating steps. It was also discovered during experimentation (see Figure 2) that pin holes in the films allowed electrochemical reactions between the metal layers to occur when submersed in NaCl solutions and caused severe device failure. To remedy this problem, platinum leader structures were used such that any sensor opening occurred only over platinum and not over Al/Pt stacks.

Second Electrode Material

One of the intended applications of these CMOS-integrated sensors was for environmental monitoring of trace metals. Using techniques known as stripping voltammetry, this type of detection is possible, but it has been shown that gold is a more suitable material for this type of analysis than platinum [11]. Consequently, a second metallization layer was introduced into the procedure to make gold electrodes available. A layer of 20/100nm Ti/Au was deposited using evaporation and was patterned using lift-off techniques. The Ti serves as an adhesion layer between the Pt and Au.

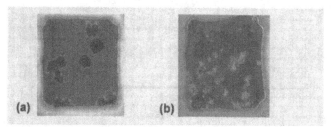

Figure 2. (a) Sputtered Al/TiN and evaporated Ti/Pt metallization stack. (b) Sputtered Al/TiN/Ti/Au metallization stack. The figures demonstrate that after soaking in a saturated NaCl solution, electrode corrosion occurs even with the TiN diffusion barrier.

Passivation

Since these transducers are submersed in solutions (often for extended periods), the passivation of the device is critical. A number of different materials that are amenable to thin-film processing were investigated by forming these materials on top of an interdigitated electrode structure. The electrode fingers were separated by 8μm and coated with a passivation layer. The passivation layer was then etched to provide vias in the field areas of the die. This configuration enabled both through-layer and lateral solution uptake to be characterized. The impedance of the electrodes were measured using an HP4145 (*Agilent Technologies, Palo Alto, CA*) gain/phase analyzer at a temperature of 95°C as the devices continuously soaked in saturated NaCl solutions. The results of the experiment are shown in Figure 3. The results clearly indicate that the spin-on polymer materials such as polyimide and parylene form poor encapsulants. Films such as silicon nitride, silicon carbide, and low-temperature oxide, on the other hand, had the best characteristics. Due to the ease of deposition and the relatively low deposition temperature of PECVD silicon nitride, this was chosen as the device's encapsulation material. The 800nm thick silicon nitride layer was deposited at 400°C. Further investigations into the silicon nitride passivation layer can be found in [7,12].

Contact Openings

Openings to the electrodes were performed using a reactive ion etch (RIE) of the silicon nitride layer. As is typical in standard CMOS RIE steps, process induced damage (PID) can occur as the electrodes charge during the etching and can permanently destroy the gates of the submicron MOSFETS. This is particularly a problem with the ISE structures as they have a large surface area. To combat PID, small reverse biased diodes like those typically used for electrostatic discharge (ESD) structures were attached to each electrode. This presented a trade-

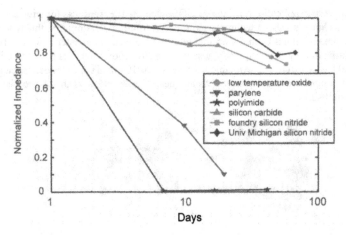

Figure 3. Normalized impedance of interdigitated electrodes versus soaktime for various encapsulation materials.

off between the capacitance and leakage current of these ESD structures versus the linearity, speed, and detection limit of the transducers themselves. The sensors employed in this work, however, did not show detrimental performance due to these added structures [8].

Third Electrode Material

The electrochemical sensors in this microsystem require reference electrodes. While much research continues in the area of reference electrodes [13], the Ag/AgCl pseudo-reference electrode is a suitable, short-term solution that is easily manufactured using thin-film silver. The 20/500nm Ti/Ag layer was sputtered in an argon atmosphere and patterned using lift-off. Care was taken in post-silver processing steps as the silver is a highly reactive material and is easily damaged in oxygen plasmas.

Polymer Containment/Exclusion Rings

During later device manufacturing steps, epoxy encapsulants and ISE membrane materials must be applied to the surface of the die in precise ways. One method for performing this application was shown in [14]. It consists of forming high-aspect ratio structures that either contain a mixture or exclude an encapsulant by surface tension. In the CMOS-integrated transducers, this was accomplished by photo-patterning an SU8 layer 13μm thick.

Back-End Manufacturing

At this point in the fabrication, the devices were coated with a thin layer of photoresist to protect the electrode surfaces until their desired use. The handling areas of the die were then diced or scribed off. The devices were then mounted onto printed circuit boards and wirebonded to make electrical connection. The wirebonds were sealed with a two-coat epoxy process. First, a low viscosity epoxy (*MED-4211, NuSil Silicon Technology, Santa Barbara, CA*) was applied and allowed to cure overnight at room temperature. Then, a coat of silicone rubber (*RTV3140, Dow Chemical, Midland, MI*) was applied over the MED-4211 and cured at 70-80°C. The silicone rubber creates a thick, high-impedance encapsulant.

The protective photoresist layer was then removed using acetone, isopropanol, and 18 MΩ deionized water. The silver electrodes were then chloridized by submerging the device into a bath of 1M FeCl₃ for 30-120s. The formation of Ag/AgCl using this technique has been problematic, hard to control, and is further aggravated by the SU8 containment rings. The rings trap air bubbles and prevent the FeCl₃ solution from reaching the silver electrodes. Research has been performed on chloridizing the silver using a plasma activated chlorine environment [7]. Under precise computer control of gas flows, temperature, and pressure, this procedure has proven to be more stable and reliable. While the current batch of devices were not fabricated using this new technique, future revisions should employ this valuable chloridization procedure.

The final fabrication step was the dispensing of a solvated ISE membrane cocktail into the containment rings. The solvent was allowed to evaporate and the hardened polymer matrix containing the ionophore for selectivity remained. Details of the membrane cocktails can be found in [4,13]. Figure 4 shows a micrograph of a completed CMOS-integrated sensor chip and a scanning electron micrograph of several ISEs.

Figure 4. Micrograph of a complete CMOS-integrated liquid chemical sensing systems and a SEM of a single ISE.

CHARACTERIZATION OF CMOS-INTEGRATED SENSORS

Voltammetric Sensors

Figure 5 shows a block diagram of some of the electronic components integrated onto the transducer array. A full description of the electronics and their characterization can be found in [8]. The electronics were initially tested before sensor post-processing, but are these results valid after post-process? To answer this question, a single NMOS transistor (W/L=10.05/1.8μm) was characterized before processing and after each individual processing step. While not a true statistical sampling, the results shown in Figure 6, indicate that there was no catastrophic failure of the device and that its extracted properties varied by less than 10%.

The CMOS-integrated voltammetric sensors were then calibrated in a solution of potassium ferricyanide using the on-chip auxiliary ($4 \times 10^{-4} cm^2$), reference ($3 \times 10^{-3} cm^2$), and working electrodes (1×10^{-4} to $1 \times 10^{-6} cm^2$). The results of this calibration were linear over a 3 decade

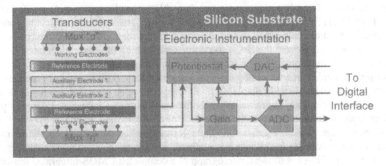

Figure 5. Block diagram of some of the electronics integrated onto the chemical microinstrument.

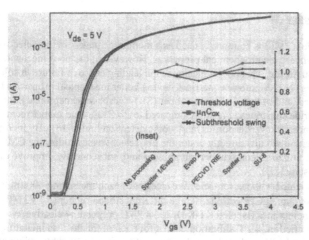

Figure 6. NMOS drain current versus gate voltage curve after each processing step. (Inset) Normalized change in extracted NMOS parameter versus processing step.

concentration range with R^2=0.99. The sensors were then used to detect trace concentrations of lead in aqueous solutions using a subtractive, square-wave anodic stripping voltammetry technique [8]. Lead was detected at 0.3ppb from two $3.2 \times 10^{-5} cm^2$ electrodes. Compared to the best detection limits achieved from our passive sensors (1.2ppb from a $4 \times 10^{-4} cm^2$ electrode), this represents a 25x improvement in detection limit/area metric.

This highly-integrated system can also provide many benefits for implantable biomedical devices. One such application is in neurochemical sensing. The neurochemical transmitter, dopamine, has been detected using the integrated system as shown in Figure 7. Characterization of the sensors for improved detection of dopamine continues.

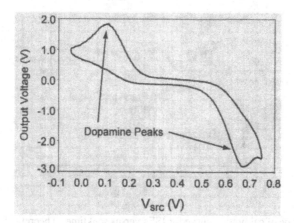

Figure 7. Cyclic voltammogram taken with the CMOS-integrated system showing the detection of dopamine.

Ion-Selective Electrodes

Poly(vinylchloride) (PVC) is commonly used as a membrane matrix in ion-selective electrodes due to its superb electrochemical properties. However, it has poor mechanical adhesion to the silicon nitride. This poor adhesion is the leading cause of failure in solid-state potentiometric sensors [16]. Membrane matrices having better mechanical properties have been previously investigated, and of these, silicone rubber (SR) has shown exceptional adhesive strength to silicon nitride substrates [17,18]. Compared to PVC, silicone rubber membranes have larger source impedances, slower responses, and worse detection limits [18]. By incorporating impedance transformation circuits directly under the ion-selective electrodes, the CMOS-integrated system can utilize silicone rubber membranes and, subsequently, improve the response and lifetime of solid-state ISEs.

Devices were fabricated with six ion-selective electrodes each measuring 400µm in diameter with a 200µm diameter Ag/AgCl electrode. Three different silicon rubber (RTV3140) potassium-selective membranes (labeled K1-K3), and a PVC potassium-selective membrane (K4) were cast as specified in [4]. Calibration curves for each of the devices including conventional macro-sized ISEs and passive solid-state ISEs were obtained by adding set amounts of KCl to a 50mM Tris-H_2SO_4, pH 7.4 buffer solution and recording the sensor response. The dynamic response of the sensors demonstrated the response time dependence on the membrane impedance and load capacitance. The CMOS-integrated ISEs responded, on average, nearly 200x faster than passive, ion-selective electrodes. They responded 7.5x faster than conventional ISEs despite having a 225x smaller membrane. The lifetimes of the ISEs were compared by running daily calibration curves. When not in operation, the ISEs continuously soaked in the buffer solution. Figure 8 shows the lifetimes of the various devices. The PVC membrane failed after 3days while the passive silicone rubber membrane survived nearly 50days before unexpectedly failing. All six of the passive ISEs failed on the same day indicating some catastrophic failure on the die. This failure is most likely unrelated to failures in the ion-selective electrodes themselves. The active ISEs with silicone rubber membranes functioned properly for more than 120days.

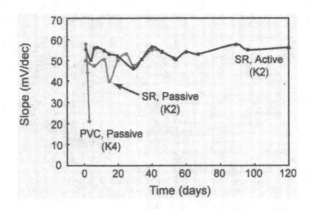

Figure 8. Slope of calibration curve for ISEs versus soaktime. Theoretically predicted slope is 59mV/dec.

CONCLUSIONS

A post-CMOS processing procedure for fabrication of CMOS-integrated liquid chemical sensors on top of foundry fabricated electronics has been detailed. The 5 mask process presents a low-cost fabrication technique for manufacturing CMOS-integrated microsystems. Two case studies were given using this process. First, a voltammetric detection system was presented for use in environmental monitoring applications and biomedical implants. The active transducers improved detection limits while decreasing device dimensions. In the second study, active ISE arrays were developed which can monitor a number of small ions important to biological functions and ecosystem health. It was shown that the CMOS-integrated devices improved response time and lifetime while again decreasing the system size. If society is to achieve pervasive sensing, low-cost microinstruments will be required to reduce the size, power, and detection limit of sensors and systems. The technologies presented here will hopefully enable a wide range of these microinstruments and have significant impact on critical applications.

ACKNOWLEDGEMENTS

The authors would like to thank Geun Sig Cha, Jeonghan Ha, Jin Wook Kim, and Arvand Salian for their help in performing some of the experiments detailed in this work. This work has been supported in part by an NSF Graduate Research Fellowship and by the Engineering Research Centers Program of the U.S. National Science Foundation under Award Number EEC-9986866. The authors would also like to acknowledge the MOSIS MEP research support program for fabrication of the CMOS test chips.

REFERENCES

1. F. Van Steenkiste, E. Lauwers, J. Suls, et. al., "A Biochemical CMOS Integrated Multi-Parameter Microsensor," *Digest of Tech. Papers Transducers 1999*, pp. 1188-1190, 1999.
2. A. Witvrouw, F. Van Steenkiste, D. Maes, et. al., *Microsystem Technologies*, **6**, 192-199 (2000).
3. A. Bard and L. Faulkner, *Electrochemical Methods*, (John Wiley & Sons, 1980).
4. S. Martin, J. Ha, J.W. Kim, et. al., "ISE Arrays with Improved Dynamic Response and Lifetime," *Tech. Digest of the Solid-State Sensor, Actuator, and Microsystems Workshop*, pp. 396-399, 2004.
5. W. Morf, *The Principles of Ion-Selective Electrodes and of Membrane Transport*, (Elsvier, 1981).
6. T. Strong, H. Cantor, and R. Brown, "A Microelectrode Array for Real-Time Neurochemical and Neuroelectrical Recording In-Vitro," *Tech. Digest of the Solid-State Sensor, Actuator, and Microsystems Workshop*, pp. 29-32, 2004.
7. T. Strong, PhD. Thesis, University of Michigan, 2004.
8. S. Martin, PhD. Thesis, University of Michigan, 2005.
9. [Online] http://www.mosis.org
10. P. Kissinger and W. Heineman, *Laboratory Techniques in Electroanalytical Chemistry 2nd Edition, Revised and Expanded*, (Marcel Dekker, Inc., 1996).

11. J. Wang, *Stripping Analysis...Principles, Instrumentation, and Applications*, (VCH Publishers, 1985).
12. G. Schmitt, J. Schultze, et. al, *Electrochimica Acta*, **44**, 3865-3883 (2000).
13. H. Nam, G.S. Cha, T. Strong, et. al., *Proc. of the IEEE*, **91**, 870-880 (2003).
14. R. Hower and R. Brown, U.S. Patent No. 6,764,652, (20 July 2004).
15. S. Martin, T. Strong, R. Brown, "Design, implementation, and verification of a CMOS-integrated chemical sensor system," *Proc. of the 2004 Intl. Conf. on MEMS, NANO and Smart Systems (ICMENS'04)*, pp. 336 - 342, 2004.
16. R. Brown, PhD. Thesis, University of Utah, 1985.
17. I. Yoon, D. Lee, H. Nam, et. al., *Jnl. of Electroanalytical Chemistry*, **464**, 135-142 (1999).
18. E. Malinowska, V. Oklejas, R.W. Hower, et. al., *Sensors and Actuators B*, **33**, 161-167 (1996).

Mater. Res. Soc. Symp. Proc. Vol. 869 © 2005 Materials Research Society · · · D3.8

A Novel Technology to Create Monolithic Instruments
for Micro Total Analysis Systems

Konstantin Seibel, Lars Schöler, Marcus Walder, Heiko Schäfer, André Schäfer, Tobias Pletzer[1],
René Püschl, Michael Waidelich, Heiko Ihmels, Dietmar Ehrhardt and Markus Böhm
Research Center for Micro- and Nanochemistry and Engineering (Cμ)
University of Siegen, D-57068 Siegen, Germany
[1] now with Institute of Semiconductor Electronics, RWTH Aachen, D-52074 Aachen, Germany

ABSTRACT

The feasibility of micro total analysis systems (μTAS) on microchips based on the concept of a monolithic instrument is demonstrated. In such a device a microfluidic layer system is deposited in a backend process on a conventional CMOS integrated circuit with the aim to achieve cost and performance enhancements through integration and miniaturization. Experimental results on elementary functional components of a μTAS are presented including a narrow channel electroosmotic micropump, a micro mass flow meter using the thermal anemometric principle, a micro cytometer with integrated optical detection, and elementary structures for on-chip microcapillary electrophoresis.

INTRODUCTION

The concept of a monolithic instrument combining standard CMOS integrated circuit technology and novel solid state components is considered a promising candidate for the realization of advanced and highly miniaturized micro total analysis systems. For that purpose a polymer based microfluidic layer system is deposited on patterned silicon wafers. The silicon wafers contain conventional integrated circuits for the control of the microfluidic components and the acquisition and processing of measurement data. A good example for an important μTAS module is a programmable micropump with an integrated mass flow sensor, measuring the actual pump rate and controlling the pump electronics in a closed feedback loop. Other examples include on-chip optical detection of fluorescence marked chemical substances using integrated photodiodes in a microcapillary electrophoresis system. The aim is an application specific lab on a microchip (ALM) of a size of a few square millimeters, at costs hardly higher than the equivalent costs for a standard CMOS chip, performing complex analytical tasks, like for example the detection of contaminants in water, e.g. Hg, or the detection of key health indicators in the human blood, e.g. K ions.

We envision mass applications for such monolithic systems mainly for environmental and life science areas, preferably one-time use applications, where a chip communicates with a PC, preferably wirelessly, and where the chip is provided similarly to and at costs comparably to, say an aspirin pill. In addition to such long term and still visionary mass applications the concept is believed to provide both enormous potential for multi-use high-end analytical applications as well as endless possibilities for fundamental research in the area of micro- and nanotechnologies, e.g. the study of chemistry in micro cavities, thanks to the direct local access to chemical / physical phenomena. While the concept and the technology platforms have been discussed elsewhere [e.g. 1, 2, 3, 4] this paper focuses on first experimental results.

EXPERIMENTAL

Micropump

Fully integrated micro total analysis systems such as a lab-on-microchip require precise transport of reagents and analytes from reservoirs to reaction chambers and separation columns. Mechanical pumps and valves have large size, are too complicated for monolithic integration on microchips and have low long-term reliability because of moving parts. The electrokinetic effect is the most common one used for nonmechanical pumping of aqueous solutions in microchannels [5]. Upon application of an electrical field E along the channel the fluid moves plug-like with velocity

$$v = \mu E = \frac{\varepsilon \varepsilon_0 \zeta}{\eta} \frac{V}{L}$$ (1)

where μ is the electroosmotic mobility, ε, ζ, η denote the dielectric constant, the surface potential and the viscosity of the fluid, respectively, ε_0 is the permittivity of vacuum, V is the voltage drop across the electrodes with the distance L. For the wide and shallow channel with a cross section of w·d the maximum flow rate is

$$Q_{max} = w d \mu E$$ (2)

at zero backpressure. The maximum pressure is achieved for zero net flow:

$$\Delta P_{max} = \frac{12 \varepsilon \varepsilon_0 \zeta V}{d^2}$$ (3)

These equations are valid as long as the Debye length λ_D of the electric double layer (EDL) is much smaller than the channel dimensions [5]. For typical aqueous solutions λ_D is smaller than 100 nm.

The main problem of electroosmotic micropumps is water electrolysis and bubble generation on metal electrodes. Therefore, for most applications, electrodes are placed into open reservoirs. Since the distance between electrodes in this scheme is large, significant pump rates require operation voltages in the kV range. Placing of electrodes within microchannels allows low-voltage pumping and enables local control of fluidic manipulations. In principle bubble generation can be suppressed by applying AC voltage over an array of interdigitated asymmetric electrodes [6] or by pulsing a zero net current signal across electrodes at low frequencies [7], but this kind of micropumps with AC activation is extremely sensitive to the hydrostatic pressure difference and cannot be used for practical applications [8]. Another possibility to eliminate the detrimental effects of gas bubbles is to place the metal electrodes outside the main flow path using gel [9] or liquid bridges [10] as ion conductors.

We have developed an electroosmotic micropump with integrated electrodes and a gas permeable poly(dimethyl)-siloxan (PDMS) cover to allow electrolysis gases to escape through. The layout of the micropump is depicted in figure 1. The active region consists of several vertical narrow microchannels with a cross section of 4 μm×15 μm to

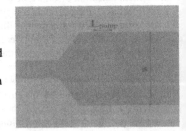

Figure 1. Layout of the micropump.

enhance pumping rate and pressure. The ribs of SU-8 are placed between gold electrodes. The distance between the electrodes is 40 μm. Figure 2 shows the SEM micrograph of the device prior to cover bonding.

The micropump was tested with DI water as working fluid, which was doped with 1 μm fluorescent latex beads. The fluid transport was monitored using an optical microscope equipped with a video camera. The maximum pumping velocity in the field-free region is about 100 μm/s corresponding to a flow rate of 5 nl/min for operating voltages of 4-6 V. The application of higher voltages results in visible gas bubbles, because the generation rate of electrolysis products exceeds the gas permeability of the PDMS cover.

Figure 2. SEM micrograph before cover bonding.

Micro Flow Sensor

Figure 3 illustrates the geometric design of the micro flow sensor (MFS). The MFS uses the anemometric principle and consists of three platinum resistors. Platinum is chosen as the sensor material because of its excellent chemical resistance and a high temperature coefficient. The heating element is located in the middle of the flow channel (50 μm×15 μm×8 mm). The resulting temperature gradient is detected by two temperature sensitive resistors according to an analytical model presented in [11]. We carried out a three-dimensional simulation based on the solution of Navier-Stokes- and energy conservation equations using the commercial simulation tool CFDRC [12].

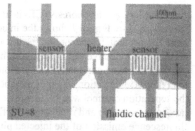

Figure 3. Photomicrograph of the MFS.

Figure 4 shows the temperature distribution at zero flow. For a heating power of 3.4 mW the maximum temperature of the heater for zero flow is 18.4 K above ambient temperature. Due to the symmetric heat dissipation along the channel, no temperature difference between the sensor elements can be observed. Figure 5 depicts the temperature distribution for a moving fluid. The isotherms are displaced in the direction of the flow. In this case a flow rate of 9 μl/min caused a temperature difference between the sensors elements of 4 K. The

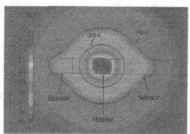

Figure 4. Temperature distribution at zero flow.

simulations assume glass as substrate material. To achieve a higher temperature difference between the sensor elements, a thermal insulation between the substrate and the fluidic channel may be introduced, e.g. SU-8. In a silicon chip, because of the high heat conductance of silicon, the electrodes need to be placed over microstructured cavities, using sacrificial layer lithography. Figure 6 shows the relationship between temperature difference and flow rate. The measurement results are in good agreement with the simulation. For flow rates up to 2 μl/min the output signal depends linearly on the flow rate in both directions. For a sensor current of 1mA and a sensor resistance of 300 Ω a sensitivity of 1.4 µV/(nl/min) can be achieved on glass with SU-8 insulation layer. For that configuration, isolation cavities are not required.

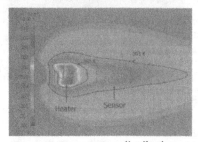

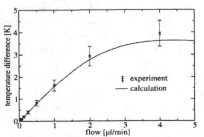

Figure 5. Temperature distribution at a flow rate of 9 µl/min.

Figure 6. Comparison between simulation and measurement.

Capillary Electrophoresis

Capillary electrophoresis (CE) is an important detection technique that should be implemented on chip. Figure 7 shows the injection cross and an electroosmotic pump for sample injection. The separation channel on the test device is designed meander like. It has a total length of 45 mm. A conductive sensor with four metal electrodes (Figure 8) is integrated at the end of this channel, allowing conductivity detection with both, two- and four-electrode setups [13]. Alternatively, optical detection schemes may be applied. To prove the elementary function of the setup the separation channel was filled with a solution of four parts ethyleneglycol and one part of borate buffer solution and rhodamine 6G fluorescence dyes. Figure 9 depicts the rhodamine 6G fluorescence emission of the injected plug. The detection volume amounts to 75 pl. Further work focuses on the optimization of the plug injection, on the design of the conductivity sensor and the reduction of plug widening by providing homogeneous zeta potential at the walls of the separation channel.

Figure 7. SEM micrograph of the CE injection cross.

Figure 8. Conductivity sensor.

Figure 9. Rhodamine 6G fluorescence plug.

Monolithic Integrated Optical Detection

Since the early 90's several groups have started to develop optics for microchips used in total analysis systems [14]. However, there is still a tremendous need for developing miniaturized fluorescence detection systems [15]. Most notably, due to the higher absorption coefficient for visible light and the low dark current an amorphous silicon detector is more suitable for the detection of fluorescence light than a crystalline silicon detector. Furthermore, most practical labelling dyes used for chemical and biological analysis target this light spectrum.

A multipurpose test structure (1.5×1.5) cm^2 depicted in Figure 10 has been fabricated. To test the lab-on-microchip configuration in chemical analysis, 2-[2'-(6'-methoxyanthryl)]-4,4-dimethyl-2-oxazoline (Ox) may be an appropriate system. The excitation light, $\lambda_{ex} = 380$ nm, enters through the thicker bottom plate and irradiates the chromophore in the channel. A fraction of both, the emitted fluorescence and the excitation light are collected in the detector. The chromophore exhibits the usual properties of anthracene derivatives, however, upon protonation a significant red shift of the absorption and emission maxima $(\lambda_{fl} = 544$ nm) takes place [16].

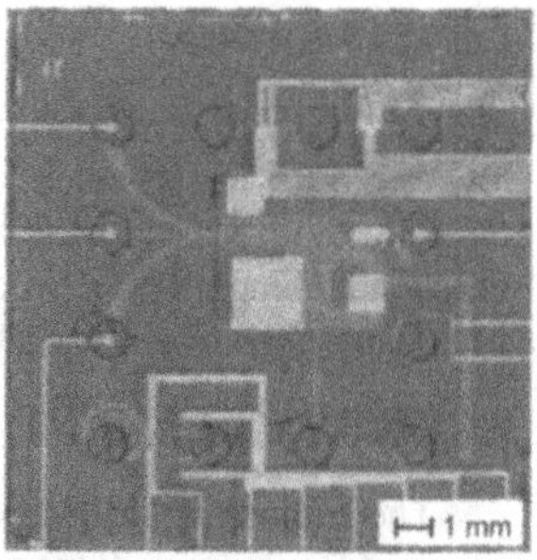

Figure 10. Photomicrograph of the fabricated multipurpose test structure.

The sample liquid was injected into the middle inlet port and the anthracenyl-oxazoline solution was pumped in a channel at a constant flow rate of ~1 µl/min. Visualisation of the hydrodynamical flow focusing results in the frame depicted in Figure 11. After initialization of the buffer and sample reservoirs with 381 mg Na$_2$B$_4$O$_7\bullet$10H$_2$O:100 ml DI-water and fluorescence dye dispersion R6G, the buffer sheath flow was driven by suction of an external syringe pump connected to the waste reservoir.

Figure 11. Hydrodynamical flow focusing in the micro cytometer.

Lab-on-microchip test results demonstrate a significant difference in the normalized intensity spectrum of Ox and Ox-H$^+$ as a consequence of the difference between the fluorescence maxima of Ox and Ox-H$^+$ (Figure 12). The diode current ranges from (23-32) pA at an illumination density of (9-12) µW/cm^2 for (400-700) nm. Experimental test parameters are bias voltage $V_{bias} = -2.5$ V, detection volume $\Delta V \sim 26$ nl, fluid concentration $c \sim 10^{-5}$ M, diode area $A_{Det} = 0.1225$ mm^2 and a constant fluid flow of ~ 1 µl/min.

The amount of substance detected was estimated to n $\approx$ 260 fmol. This is comparable to the results obtained by Fixe et al. [17a] who specified a detection range of 2.5-700 pmol DNA/cm^2 in a two dimensional configuration for DNA analysis with a photoconductor based on amorphous silicon, an interference filter composed of 15 layers deposited by SiO$_2$, SiN$_x$ and an a-SiC:H UV light filter. Referring to the metal contact geometry used by Fixe et al. consisting of two pads with 600 µm length, 200 µm width [16b] and 100 µm distance between, the total amount of substance could be calculated to the range of 7.5-2100 fmol.

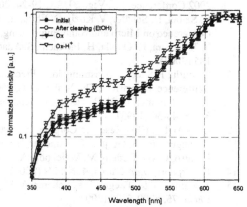

Figure 12. Normalized light intensity in initial phase, at runtime and after cleaning.

CONCLUSION

The concept of a monolithic instrument is considered a promising approach to miniaturized low-cost high-performance total analysis systems providing enough room for specific developments for both mass market and high-end applications. Experimental results presented on various functional components of such systems, like pumps, flow sensors and optical detection systems demonstrate the feasibility as well as the need for further fundamental research.

ACKNOWLEDGEMENT

The authors would like to thank the staff of the Research Center for Micro- and Nanochemistry and Engineering (Cμ) at University of Siegen. This work is funded by the Deutsche Forschungsgesellschaft (DFG), project BO 772.

REFERENCES

1. H. Schäfer, S. Chemnitz, K. Seibel, V. Kozij, A. Fischer, D. Ehrhardt, M. Böhm, in *The Nano-Micro Interface*, eds. H.-J. Fecht, M. Werner (WILEY-VCH Verlag, 2004), p. 119.

2. H. Schäfer, K. Seibel, M. Walder, L. Schöler, T. Pletzer, M. Waidelich, H. Ihmels, M. Schmittel, D. Ehrhardt, M. Böhm, Proc. Micro Total Analysis Systems 2004, vol.2, 443-445, 2004.

3. H. Schäfer, K. Seibel, M. Walder, L. Schöler, T. Pletzer, M. Waidelich, H. Ihmels, D. Ehrhardt, M. Böhm, Proc. Micro Electro Mechanical Systems 2005, 758-761, 2005.

4. L. Schöler, B. Lange, K. Seibel, H. Schäfer, M. Walder, N. Friedrich, D. Ehrhardt, F. Schönfeld, G. Zech, M. Böhm, 30th International Conference on Micro and Nano Engineering, September 19-22, Rotterdam, NL, 2004, Microelectronic Engineering, in press, 2004.

5. C.-H. Chen, J. Santiago, *Journal of Microelectromechanical Systems* Vol. 11, No. 6, 672-683 (2002).

6. V. Studer, A. Pépin, Y. Chen, A. Ajdari, *Microelectronic Engineering* 61-62, 915 (2002).

7. S. Mutlu, C. Yu, P. Selvaganapathy, F. Svec, C. Mastrangelo, J. Fréchet, Proc. IEEE MEMS 2002 Conference, Las Vegas, USA, 19-24, 2002.

8. K. Seibel, H. Schäfer, V. Kozij, D. Ehrhardt, M. Böhm, presented at the 29th International Conference on Micro and Nano Engineering, September 22-25, Cambridge, UK, 2003.

9. Y. Takamura, H. Onoda, H. Inokuchi, S. Adachi, A. Oki, Y. Horiike, *Electrophoresis* 24, 185 (2003).

10. S. Mutlu, F. Svec, C. Mastrangelo, J. Fréchet, Y. Gianchandani, Proc. IEEE MEMS 2004 Conference, Maastricht, The Netherlands, 850-853, 2004.

11. T. S. J. Lammerink, N. R. Tas, M. Elwenspoek, J. H. J. Fluitman, *Sensors and Actuators A* 37-38, 45-50 (1993).

12. CFD-ACE+, CFD Research Corporation, 2004, Huntsville, USA, (http://www.cfdrc.com).

13. F. Laugere, R. M. Guijt, J. Bastemeijer, G. van der Steen, A. Berthold, E. Baltussen, P. Sarro, G. van Dedem, M. Vellekoop, A. Bossche, *Anal. Chem.* 75, 306-312 (2003).

14. E. Verpoorte, *Lab Chip* 3, 42N-52N (2003).

15. T. Kamei, B. M. Paegel, J. R. Scherer, A. M. Skelley, R. A. Street, R. A. Mathies, *Anal. Chem.* 75, 5300-5305 (2003).

16. H. Ihmels, A. Meiswinkel, C. J. Mohrschladt, *Org. Lett.* 2, 2865 (2000).

17. a) F. Fixe, V. Chu, D.M.F. Prazeres, J.P. Conde, *Nucl. Acid Res.* 32, e70 (2004).
    b) V. Chu (private communication).

Functional Oxides and Other Materials
for Monolithic Instrument Integration

Mater. Res. Soc. Symp. Proc. Vol. 869 © 2005 Materials Research Society   D1.7

Integration of Zinc Oxide Thin Films with Polyimide-based Structures

Masashi Matsumura[1], Zvonimir Bandic[2], and Renato P. Camata[1]
[1] Department of Physics, University of Alabama at Birmingham,
Birmingham, AL 35294, U.S.A.
[2] Hitachi Global Storage Technologies, Inc., San Jose, CA 95120, U.S.A.

ABSTRACT

Zinc oxide (ZnO) films were deposited on polyimide-based substrates by pulsed laser deposition. The ZnO films and the underlying polymer layers were studied using Fourier Transform Infrared (FTIR) and Photoluminescence (PL) spectroscopies. FTIR measurements in structurally well-characterized samples exhibiting all X-ray reflections of crystalline hexagonal ZnO show absorbance bands around 405 cm^{-1} (Zn-O stretching vibration) and 1110 cm^{-1} (in-plane C-H vibrations on aromatic rings of polyimide). Observed shifts in both absorption features as a function of deposition temperatures can probably be attributed to thermal stress in the layers. PL measurements showed broad spectra centered around 3.35 eV from ZnO excitonic emission and a broader PL band from 2.7 to 3.1 eV defect complex emissions. The appearance of these peaks was consistent for depositions of ZnO on non-organic substrates, which indicates that the integration of ZnO thin films on polymer based substrate preserved the characteristic optical properties of ZnO.

INTRODUCTION

ZnO has a wide band gap of 3.37 eV at room temperature, and has large exciton binding energy (59 meV), which shows excitonic effects even at room temperature. With its hexagonal wurtzite structure, ZnO is piezoelectric [1], has ferroelectric properties when doped [2], and often exhibits n-type conductivity [3]. By taking advantage of these characteristics, ZnO is widely used in various technological applications such as surface acoustic wave devices, pyroelectric sensors, piezoelectric devices, gas sensors, and varistors among others. ZnO is particularly attractive for monolithic instruments because it is relatively easy to obtain good quality polycrystalline films with moderate Hall mobilities (>1 cm^2/V s) at room temperature [4]. This affords this material good compatibility with plastic or flexible substrate materials. Since ZnO is transparent in the visible, it is also compatible with a variety of applications in flexible photovoltaics [5]. ZnO has attracted attention not only as a thin film semiconductor material, but also in exquisite nanoscale configurations [6]. This is a result of the recent demonstration of numerous methods for ZnO nanostructure synthesis. Several developments remain to be achieved, however, if ZnO is to become a reliable commercial semiconductor material. These include advances in bulk and epitaxial growth [7], p-type doping [8], and high quality integration to polymer platforms. In this paper a study of the microstructural and optical properties of ZnO films integrated onto polymer-based substrates under different film deposition conditions was performed in order to evaluate the possibility of future monolithic devices that combine the properties of ZnO films and polymer substrates.

EXPERIMENTAL DETAILS

ZnO films on polyimide substrates were obtained using a KrF excimer laser (248 nm) with energy density of 3.0–6.2 J/cm^2 to ablate a hot-pressed ZnO powder ceramic target. During depositions, a pulsed laser deposition (PLD) system was operated at a pressure of 0.1 to 50 Pa of O$_2$ gas (99.999% purity) and the substrate temperature was varied from room temperature to 400 °C. We have used two types of polyimide substrates.

In both cases the polyimide biphenyl 1,1'-2,2'-tetracarboxydianhydride/ 1,4-phenylenediamine (PI-2611, HD MicroSystems, Cupertino, CA, USA) was employed. The chemical structure of this polyimide is shown in Fig. 1. In one set of samples the substrates were prepared by spin coating clean oxidized silicon wafers with a thin layer of the PI-2611 precursor followed by curing which resulted in an adherent polyimide films of about 5-μm thickness on the wafers. ZnO deposition was then performed by PLD. A second set of samples was produced by depositing ZnO on free-standing polyimide films with a typical thickness of 15 μm. These films were prepared by spin coating clean oxidized silicon wafers with a thin layer of the PI-2611 precursor followed by curing and wafer separation. Through the laser deposition process, ablation species with high kinetic energy are delivered either to the polyimide/Si substrate *or* free-standing polyimide film surface to form ZnO films. On both types of substrates we have deposited films with nominal thickness between 100 nm and 9.0 μm as determined from calibrations of the deposition rate using surface profilometry measurements. Because of the marked difference between the coefficient of thermal expansion of ZnO (4.75 × 10^{-6} /K) [9] and that of polyimide

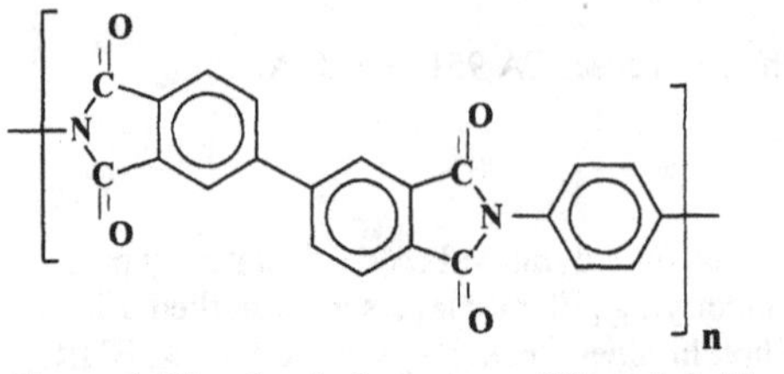

Figure 1. The chemical structure of the polyimide biphenyl 1,1'-2,2' tetracarboxydianhydride/1,4-phenylenediamine used in this study (PI-2611, HD Microsystems).

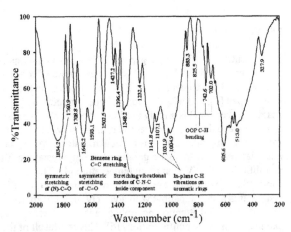

Figure 2. Fourier Transform Infrared Absorption of the polyimide PI-2611. Of particular interest are the in-plane C-H vibrations of aromatic rings in the 1000-1100 cm^{-1} wavenumber range.

$(30 \times 10^{-6}/\mathrm{K})$, we anticipated the development of significant thermal stresses particularly in films deposited at elevated temperatures. Fourier Transform Infrared Spectroscopy (FTIR) provides a simple and sensitive way to monitor the effect of these stresses on the structural properties of samples, which result from the integration of such dissimilar materials. Selected vibrational frequencies of ZnO and polyimide may be particularly sensitive to strain fields in the films and substrates allowing the assessment of thermal stresses and correlations with other properties. Fig. 2 shows the multiple absorption bands of the polyimide used in our studies that could be affected by intrinsic and thermal stresses generated as a result of the deposition. Using the spectrum in Fig. 2 as a reference, we performed FTIR measurements on samples containing ZnO films deposited at various temperatures under a pressure of 0.1 Pa. A Bruker Tensor 27 FTIR spectrometer working at a resolution of 4 cm^{-1} was used for these measurements. In order to establish the crystallinity of these ceramic ZnO films deposited on the polymeric substrates, our films were also analyzed by X-ray diffraction (XRD) using a thin film diffractometer. Photoluminescence (PL) scans were carried out with a Fluoromax 2 fluorometer with excitation carried out using photons of 3.875 eV (320 nm). Additional experimental details are provided in the next section.

DISCUSSION

ZnO films deposited on polyimide layers on silicon (polyimide/Si)

Figure 3. Schematic of layered structure of ZnO films on polyimide/Si for basic studies.

Although our ultimate goal is the integration of ZnO films onto flexible polymer platforms, rigid samples of polyimide films spun on Si wafers provide very well characterized substrates for basic studies (Fig. 3). As expected, the thermal stress in our hybrid films can build to very significant values leading to stress relaxation through cracking and delamination as evidenced by the optical micrograph in Fig. 4(a) for a ZnO film with a thickness of 1.2 µm. We have therefore concentrated our depositions on films below the critical thickness for stress relaxation in this

system, which was empirically found to be approximately 300 nm. Fig. 4(b) shows an optical micrograph for a 100-nm thick ZnO film deposited under the same conditions, which is crack free and for which AFM analysis [Fig. 4(b) inset] yields a very smooth surface. We have subjected these crack-free smooth samples to FTIR analysis in an attenuated total reflection configuration. Fig. 5 shows the absorption spectra obtained from our samples deposited at various temperatures ranging from room temperature to 400 °C in for wavenumbers

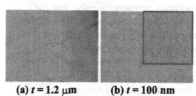

(a) $t = 1.2$ µm   **(b)** $t = 100$ nm

Figure 4. Optical micrographs of ZnO films of two different thicknesses deposited on polyimide/Si substrates. Inset in part (b) shows a 1-µm atomic force microscopy scan of the smooth ZnO film on polyimide/Si.

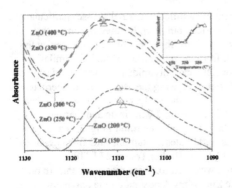

Figure 5. FTIR spectra of samples consisting of ZnO films deposited on polyimide/silicon substrates at different temperatures. A laser energy density of 2.0 J/cm² and a deposition pressure of 0.1 Pa were used. The triangles indicate the peak position of the main absorption feature.

around 1100 cm⁻¹. As previously shown in Fig. 2, this spectral region is populated by several modes of in-plane C-H vibrations on the aromatic rings of the polyimide material. Indeed, the spectra in Fig. 5 shows a very distinct absorption feature centered around 1110 cm⁻¹, which does not change substantially in character with the change in ZnO deposition temperature. In concordance with the reference data for our polyimide (Fig. 2), we attribute this absorption band to an in-plane C-H vibration in the aromatic rings of the PI-2611. As seen in Fig. 5 and plotted in its inset, a monotonic shift in the position of this C-H vibration toward higher wavenumbers is observed for samples with higher deposition temperatures. The cause for this shift may be the increased thermal stress experienced by the polyimide substrate as a result of higher deposition temperatures [10,11]. The absorption spectra in Fig. 6 shows an interesting complementary trend for an absorbance band centered around 405 cm⁻¹ and unambiguously attributed to ZnO. The FTIR spectrum for polyimide/Si substrate itself has also been included for comparison and is essentially featureless in this spectral

range [trace(a)]. The 405 cm⁻¹ band corresponds to the Zn-O stretching vibration for a tetrahedral surrounding the zinc atoms [12]. The absorption of infrared radiation at this wavenumber increases as the deposition temperature of the ZnO film increases. This indicates, not surprisingly, that ZnO films of better quality and well-defined crystalline vibrational modes are obtained at higher deposition temperatures. It is interesting to note, however, that the peak of this absorption feature appears to shift to lower wavenumbers as the deposition temperature increases. This shift may also be a result of increased thermal stress experienced now by the ZnO film due to higher deposition temperatures [10,11].

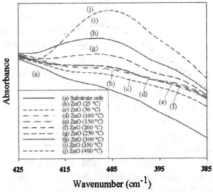

Figure 6. FTIR spectra of ZnO films deposited on polyimide/silicon substrate at different temperatures. A laser energy density of 2.0 J/cm² and a deposition pressure of 0.1 Pa were used.

ZnO films deposited on free-standing polyimide films

Figure 7. Schematic of ZnO films on 15-μm thick free-standing polyimide substrate.

Studies performed on films deposited on rigid substrates may provide important baseline information on the critical thickness and stress behavior of ZnO films on polyimide. Ultimately, however, we are interested in demonstrating functional ZnO films that can be integrated with flexible layers. We have performed, therefore, depositions and measurements on free-standing flexible polyimide substrates with a typical thickness of 15 μm (Fig. 7). These samples evidently tolerate larger thickness values of the ZnO film without cracking and delamination because of substrate flexibility. Highly flexible and adherent ZnO layers were obtained with a thickness around 100 nm as shown in Fig. 8. PL measurements on similar samples reveal a trend very similar to that observed in ZnO films deposited on other substrates. Figure 8 shows room-temperature steady-state PL spectra of samples for which ZnO was deposited at temperatures between 100 and 300 °C using a laser energy density of 3.0 J/cm^2. The insets show intensity ratios obtained from the spectra by dividing the integrated luminescence intensity in the wavelength range noted in each inset by the intensity of the 3.35-eV peak. All three spectra exhibit a broad spectral feature centered at about 3.35 eV, which results from excitonic emission in ZnO[11]. Even broader PL bands are observed between 2.5 and 3.1 eV due to defect complex emissions. This very similar PL behavior for the samples deposited between 100 and 300 °C correlates well with their comparable crystal quality revealed by X-ray diffraction measurements (not shown). As expected, an increase in the substrate temperature leads to an overall reduction in the defect-related luminescence in the 2.5–3.1-eV range compared to excitonic recombination around the band gap energy (~3.35-eV). Some subtle differences exist however in the PL bands observed from these films at 2.5–3.1 eV. We note that, in the 2.7–3.1-eV spectral region, the PL intensity seems to decrease linearly with increasing temperature. This can be seen by taking the ratio of the integrated PL intensity in the 2.7–3.1-eV range to the intensity of the 3.35-eV peak as shown in Fig. 9, inset (a). On the other hand, the reduction of PL intensity in the 2.5–2.7-eV range with increasing temperature seems to deviate significantly from linear behavior as seen in inset (b). This suggests that at least two optically active defect complexes are present in these ZnO films deposited on polyimide producing emission in two separate bands in the 2.5–3.1-eV range [13].

CONCLUSION

We have deposited and analyzed samples comprising ZnO thin films integrated with polyimide-based substrates. FTIR measurements as a function of ZnO

Figure 8. Tweezers flexing a ZnO film with a thickness of 120-nm deposited at 300°C on a free-standing, flexible polyimide film of 15-μm thickness.

deposition temperature on smooth, crack-free ZnO films on polyimide/Si detected complementary shifts in the Zn-O stretching vibration absorption of ZnO crystal near 405 cm^{-1} and in the in-plane C-H vibration on aromatic rings of polyimide PI-2611 near1110 cm^{-1}. While the Zn-O mode shifts toward lower wavenumbers, the C-H moves to higher values. The temperature dependence of such shifts suggests they are caused by thermal stress in the layers. PL measurements on similar ZnO films deposited on free-standing flexible polyimide substrates show temperature behavior reminiscent of ZnO deposited on conventional inorganic substrates such as sapphire and silicon. Higher substrate temperature, up to 300 °C, produces films with less defect luminescence.

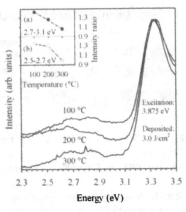

Figure 9. Room temperature steady-state photoluminescence spectra of ZnO thin films deposited on polyimide substrates at different deposition temperatures using a laser

ACKNOWLEDGMENTS

The authors would like to thank Dr. Thomas M. Nordlund for the use of the spectroscopy facilities in his laboratory as well as Dr. Andrei Stanishevsky for assistance with the FTIR measurements. We also thank Dr. Douglas B. Shire, Dr. Carmen Scholz, and Ms. Robin Sweitzer for providing the free-standing polyimide substrates used in this study. Support for instrumentation used in this research was provided by the National Science Foundation under grants DMR-0116098 (MRI - Major Research Instrumentation) and DBI-0070220 (MUE - Multi-User Biological Equipment and Instrumentation).

REFERENCES

1. M. Kadota, and M. Minakata, *IEEE Trans. Ultrason. Ferroelectr. Freq.Control* **42**, 345 (1995).
2. M. Joseph, H. Tabata, and T. Kawai, *Appl. Phys. Lett.*, **74**, 2534 (1999).
3. M. Hiramatsu, K. Imawda, N. Horio, and M. Nawata, *J. Vac. Sci. Technol.*, **A16**, 669 (1998).
4. W.-J. Jeong and G.-C. Park, *Sol. Energy Mater. Sol. Cells* **65**, 37 (2001).
5. K. Matsubara, P. Fons, K. Iwata, A. Yamada, K. Sakurai, H. Tampo, S. Niki, *Thin Solid Films*, **431** 369 (2003).
6. P. Hammer, M. S. Tokumoto, C. V. Santilli, S. H. Pulcinelli, A. F. Craievich, and A. Smith, *J. Appl. Crystallography* **36**, 435 (2003).
7. M. Ohtomo, K. Tamura, K. Saikusa, K. Takahashi, T. Makino, Y. Segawa, H. Koinuma, and M. Kawasaki, *Appl. Phys. Lett.* **75**, 2635 (1999).
8. G. Xiong, J. Wilkinson, B. Mischuck, S. Tüzemen, K. B. Ucer, and R. T. Williams, *Appl. Phys. Lett.* **80**, 1195 (2002).
9. K. Ellmer, J. Phys. D: *Appl. Phys.* **33**, R17 (2000).
10. A. Klett, R. Freudenstein, M. F. Plass, W. Kulisch, *Surf. and Coatings Tech.,* **125**, 190 (2000).
11. V. G. Gregoriu, G. Kandilioti, and K. G. Gatos, *Vibrational Spec.,* **34**, 47 (2004).
12. P. Tarte, *Spect. Chemim. Acta* **18**, 467 (1961).
13. M. Matsumura, R. P. Camata, *Thin Solid Films* **476**, 317 (2005).

Mater. Res. Soc. Symp. Proc. Vol. 869 © 2005 Materials Research Society   D2.4

RF HOLLOW CATHODE PLASMA JET DEPOPSITION OF BAxSR1-xTI03 FILMS

N.J. IANNO[1], R.J. SOUKUP[1], Z. Hubička[2], J. Olejníček[2], and H. Šíchová[2]

[1] Department of Electrical Engineering, 209N WSEC, University of Nebraska-Lincoln, NE 68588, USA; and 2 Institute of Physics ASCR, Na Slovance [2] 182 21 Prague 8 Czech Republic

Abstract

An initial study of the RF hollow cathode plasma jet deposition of $Ba_xSr_{1-x}TiO_3$ has been performed. Deposition occurred from a single composite nozzle consisting of $BaTiO_3$ and SrTiO3 at substrate temperatures on the 500-550 C range. It has been shown that film composition can be easily controlled by the nozzle composition as well as other deposition parameters. The as-deposited films exhibit clear BSTO peaks with grain size on the order of 30nm.

Introduction

Ferroelectric thin films have attracted considerable attention since they have application in dynamic random access memories (DRAM), non volatile memories, electro-optic switches, and most recently tunable phase shifters.[1-4] Barium strontium titanate (BST) combines the merits of the high permitivity of $BaTiO_3$ with the structural stability of $SrTiO_3$ and therefore is an excellent choice for the above applications.[5]

We have developed a new technique for the deposition of high quality semiconductor material, and in this work we apply it to the deposition of BST.[6-8] This technique is a hollow cathode plasma jet deposition. It has been used by others to deposit TiN, Ge_3N_4, $LiCoO_x$ and most recently we have improved the method to yield device quality a-Si:H, and a-SiGe:H. [6-12] We have deposited BST via this technique on bare silicon and Pt coated silicon wafers. We will report on the film composition as a function of the deposition conditions and post deposition processing.

Experimental Apparatus

The low pressure plasma jet configuration for BST thin film deposition can be seen in Fig 1. The reactor chamber was continuously pumped. A cylindrical composite nozzle consisting of $SrTiO_3$ and $BaTiO_3$ is mounted inside a water cooled copper block. This copper block is also the electrical contact to an RF source operating at 13.56MHz. The working gas is $Ar+O_2$ which flows through the nozzle and is controlled by electronic mass flow controllers. When driven by the RF source an intense hollow cathode discharge was generated inside the nozzle which reactively sputters the nozzle material.[13] The flowing working gas forces the plasma and sputtered material out into the chamber where deposition occurs on the substrate. The internal diameter of the nozzle was 3 mm and the total length was 30 mm. In order to deposit $Ba_xSr_{1-x}TiO_3$ thin films the nozzle was composed from two parts as can be seen in Fig 1. Since it was known that STO has a lower sputtering rate than BTO, the STO portion was placed at the nozzle outlet where hollow cathode plasma has the highest density. Previous work has shown the composition of the deposited material is a strong function of the length ratios.[14]

Based on those results the length of the STO portion was set at 25 mm while the BTO portion was set at 5 mm. The substrate was placed perpendicularly to the plasma jet axis 20 mm from the hollow cathode outlet. Films were deposited simultaneously on Si wafers and on the Pt top layer of multi-layer $Si/SiO_2/TiO_2/Pt$ substrate.

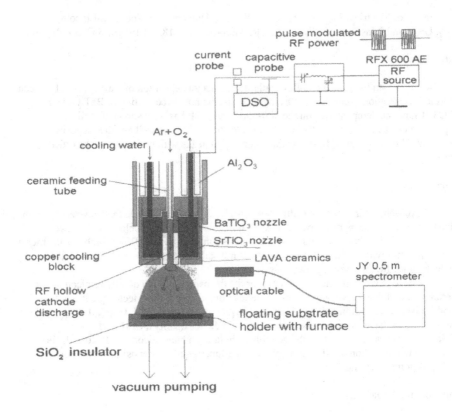

Figure 1. Experimental apparatus for the plasma jet deposition of BSTO thin films.

The analysis of the film structure was performed by x-ray diffraction in the grazing incidence(GI) geometry using CuK_α radiation. Chemical composition of deposited films was measured by electron microprobe analysis. The energy of the primary electrons in the beam was set to 10 keV. Electron micro-probe analysis provides information about the average composition over the whole profile of the analyzed films.

The current (Rogowski coil) and voltage probes with calibrated RF amplitude response and phase shift up to 100 MHz in combination with a digital oscillograph were used to measure the RF power, RF current and RF voltage on the nozzle. The temperature of the substrate was held between 500-550 °C and the total gas pressure in the reactor was on 6 Pa for all the results presented here. Other reactor conditions are listed in table 1. The RF power, P_{RF}, absorbed in the plasma was calculated from RF current and voltage waveforms together with their phase shift as measured by the digital oscillograph. The RF power, RF voltage amplitude and current measured on the nozzle electrode are presented in table 1. Emission spectroscopy was used to monitor the sputtered species contained in the plasma jet channel. The full experimental set up can be seen in Figure 1.

Table I. Deposition conditions for deposited thin films

Sample no.	Q_{Ar} [sccm]	Q_{O2} [sccm]	P_{RF} [W]	Used nozzle	I_{RFm} [A]	U_{RFm} [V]
2	85	64	219	STO,BTO	4.11	232
3	42	32	210	STO,BTO	4.1	229
4	42	13	180	STO,BTO	4.0	217
5	42	64	190	STO,BTO	3.9	220

Results

It was possible to deposit $Ba_xSr_{1-x}TiO_3$ thin films on Si and Pt surfaces. The maximum growth rate achieved was 250-300 nm/hour. Table 2 shows the composition as determined by electron microprobe analysis. As seen in the table, $Ba_xSr_{1-x}TiO_3$ films with x ranging from 0.56 to 0.44 where obtained. All the BSTO samples have some Ti excess Ti/(Ba+Sr) $\approx$ 22/17 and some oxygen excess. In order to perform a detailed study of the crystalline structure, sample 2 was analyzed by x-ray diffrction in the GI geometry (parallel beam). This method is more sensitive because for small angles between the x-ray beam and the plane of the substrate surface the larger volume of the film material is responsible for diffraction. Furthermore, the influence of texture is less significant in this set up. Several diffraction peaks belonging to the BST perovskite phase were found as it can be seen in Fig 2. From the position of the detected BST diffraction peaks, it was possible to calculate the lattice parameter a = 3.980 Å. Furthermore, it was possible to estimate the grain size of crystallites from the width of the diffraction peaks. The grain size is estimated to be on the order of *30 nm*. On the other hand this estimation is valid only in the case where we can neglect the internal stress in the films which can exist in our case.

Table II. Chemical composition of deposited samples

Sample Number	Thickness (nm)	Atomic% (O)	Atomic%(Ti)	Atomic%(Sr)	Atomic%(Ba)
2	823	61	22	7.4	9.1
3	286	61	22	7.6	9.25
4	232	60	21.5	9.2	8.3
5	216	61	22	9.2	7.3

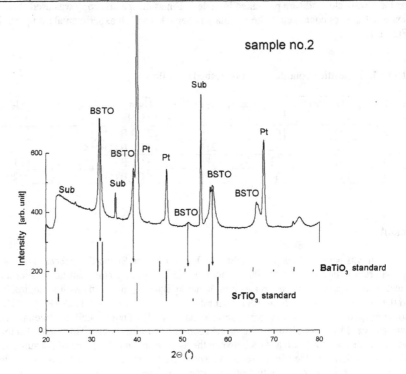

Figure 2. X-ray diffraction pattern of the sample 2 in the GI geometry.

The typical emission spectra from the plasma at the outlet of the nozzle shows Ba^+, Ba and Sr emission lines. The intensities of these lines were roughly proportional to the deposition rate. By permanent observation of those atomic lines it may be possible to directly control and reproduce the film composition. As seen in Figure 3 the ratio of atomic concentration of Ba and Sr contained in the films was approximately proportional to the ratio of emission intensity of Sr and Ba excited atoms and ions.

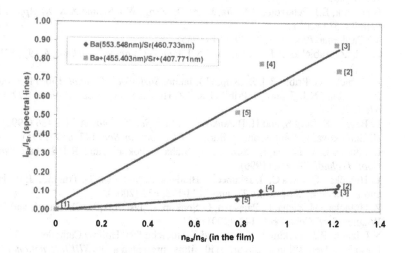

Figure 3. The ratio of the emission intensity of Ba^+/Sr^+ and the emission intensity of Ba/Sr lines versus the atomic ratio of Ba and Sr contained in the films.

Conclusion

We have the demonstrated the deposition of BST films from a composite nozzle via the hollow cathode rf plasma jet deposition technique. The as-deposited films at substrate temperatures in the 500 to 550 C range exhibit clear BST x-ray diffraction peaks.

Future work will focus on evaluating the films electronic properties and on studying the effect of post deposition annealing.

Acknowledgements

This is work is supported by ARL Grant # W911NF-04-2-0011(UNL) and junior grant KJB1010302 agency ASCR.

References

1. J.J. Scott, *Ferroelectric* **183**, 51 (1996).
2. J.F. Scott, *Integrated Ferroelectrics* **20**, 15 (1998).
3. J. Li, F. Duewaer, C. Gao, H. Chang, X.D. Xiang and Y. Lu Y, *Appl. Phys. Lett.* **76**, 769 (2000).
4. B.H. Park, E.J. Peterson, Q.X. Jia, J. Lee, X. Zeng, W.I. Si, and X.X. Xi, *Appl. Phys. Lett.* **78**, 533 (2001).
5. M. Cardonna, *Phys. Rev. A* **140**, 651 (1965).
6. Pribil, Z. Hubička, R.J. Soukup, and N.J. Ianno, *J. Vac. Sci. Technol. A* **19**, 1571 (2001).
7. Z. Hubička, G. Pribil, R.J. Soukup, N.J. Ianno, *Surf. Coat. Technol.* **160**, 114 (2002).
8. R.J. Soukup, N.J. Ianno, G. Pribil G, and Z. Hubicka, *Surf. Coat. Technol.* **177-178**, 676 (2004).
9. L. Bárdoš, S. Berg S, and H. Baránková, *J. Vac. Sci. Technol. A* **11**, 1486 (1993).
10. H. Baránková, L. Bárdoš, and S. Berg, *J. Electrochem. Soc.* **142**, 883 (1995).
11. L. Soukup L, V. Perina, L. Jastrabik, M. Šícha, P. Pokorny, and R.J. Soukup, *Surf. Coat. Technol.* **78**, 280 (1996).
12. Z. Hubička, C. Cada C, I. Jakubec, J. Bludska, Z. Malkovs, B. Trunmd, J. Pridal, and L. Jastrabik, *Surf. Coat. Technol.* **174-175**, 632 (2003).
13. Z. Hubička, M. Šícha, L. Pajasova, L. Soukup, L. Jastrabík, D. Chvostová, and T. Wagner, *Surf. Coat. Tech.* **142**, (2001) 681.
14. N.J. Ianno, R.J. Soukup, N. Lauer, and Z. Hubicka "RF Hollow Cathode Plasma Jet Deposition of $Ba_xSr_{1-x}TiO_3$ films" presented at the *XIII International Materials Research Congress 2004*, August 2004 Cancun Mexico.

Mater. Res. Soc. Symp. Proc. Vol. 869 © 2005 Materials Research Society   D2.9

Low temperature deposition of Indium tin oxide (ITO) films on plastic substrates

Vandana Singh, B. Saswat and Satyendra Kumar

Samtel Center for Display Technologies
Indian Institute of Technology Kanpur, India
Email: vandanas@iitk.ac.in

ABSTRACT

Organic light emitting diodes (OLEDs) require a transparent conducting oxide (TCO) electrode for injection of charge carriers and the emitted light to come out. In order to exploit the full flexibility of organic semiconductor based large area electronic devices, the deposition of TCO on plastic substrates is essential, which prohibits high temperature processing. Therefore, low temperature deposition of Indium tin oxide (ITO) films is very important for flat panel displays and solar cells. Here we have carried out a systematic study of ITO deposition on plastic substrates using RF magnetron sputtering. For the optimization of structural, electrical and optical properties of ITO, various experiments such as X-ray diffractometer, transmission measurements, sheet resistance and AFM were employed. These properties were investigated as a function of substrate temperature, deposition time and RF power. From these experiments, we obtained a reasonably low sheet resistance (~14 Ω /□) and high transmittance (~75%) in the visible region on plastic substrates. We also observed that these films are not much affected by atmosphere and does not degrade with time. These ITO films deposited by RF magnetron sputtering on plastic substrates can be use as anode for flexible organic light emitting displays.

Key words: ITO, OLED, Plastic Substrate and Magnetron Sputtering.

INTRODUCTION

Discovery of electro-luminescence from organic polymer poly (1,4-phenylene vinylene) PPV [1], provided an opportunity to make organic light emitting devices (OLED). These devices are lightweight, flexible and easy to fabricate on large area. Basic structure of an OLED is one or two organic layers (100nm) sandwiched between two electrodes, one of which is transparent. So, any OLED needs a thin film of transparent conducting oxide (TCO) material for injection of charge carriers and extraction of emitted light. This thin film of TCO should have low resistivity, high transmission in the visible range and high work function. ITO, a highly degenerate wide band gap semiconductor and has low resistivity (10^{-4} Ω cm), high transmission in the visible range (~98%) and high work function (4.6 ev) [2], is most widely used TCO material. However, glass is very brittle, too heavy for large area displays and cannot be deformed. These problems can be overcome by using plastic substrates, which are lightweight, robust and flexible. So plastic substrates have been used in polymeric and molecular OLEDs. For the full flexibility of organic light emitting displays, OLEDs need to be fabricated on plastic (polymer) substrate. For flexible OLED (FOLED), ITO is deposited on plastic substrates, which has relatively high resistivity compared to glass and low glass transition temperature (Tg). Due to temperature constraints, low temperature process for deposition of ITO on plastic is required.

There are various ways to deposit ITO on glass such as electron beam evaporation, ion assisted deposition [3], CVD [4], thermal evaporation [5], laser ablation [6] and magnetron sputtering [7-9]. Sputtering method is commonly preferred due to good homogeneity, good film adhesion and high purity [10].

Various efforts have been made for deposition of ITO material, but mostly on glass substrates [11]. Some papers also report deposition on polymer substrate [12]. For device application, the polymeric substrate must be transparent and thermally stable so that it does not deform under typical deposition process. Majority of studies have been done on acrylic, polycarbonate and polyethylene terephthalate (PET). In order to get ITO film on other plastic substrates with properties compatible for FOLED fabrication, detailed study of ITO film deposition is required.

In this paper, electrical, optical and nanostructural properties of ITO thin film deposited on Mylar rubylith, polycarbonate and glass substrate have been reported.

EXPERIMENTAL PROCEDURE

ITO thin films were deposited on Mylar, polycarbonate and glass by RF magnetron sputtering. The ITO target used was In_2O_3 having 10% SnO_2. The distance between target and substrate was approximately 6cm. All the substrates were ultrasonically cleaned in isopropanol and dried. Before deposition, all the samples were annealed into microwave oven at 80^0C for 30 min. The deposition process was carried out at 15 sccm of Argon gas, which was controlled through mass flow-meter. The base pressure of sputtering system was 3×10^{-6} mbar and process pressure was 2.5×10^{-3} mbar. The applied RF power was 60 W and constant during all the process. The deposition was done at different temperature for different deposition time. The maximum substrate temperature was maintained around 100^0C to avoid thermal distortion of the substrate. Plasma treatment was done on ITO coated plastic substrate for 10 minute at 15 Watts. Finally, optical, electrical and structural properties of the ITO films were studied. Transmission was measured with Perkin-elmer-UV/VIS Lambda 40 spectrometer; work function measurement was done with Kelvin probe; sheet resistance was measured with 4 point probe; surface profilometer was used for thickness and roughness measurements and XRD was used for nanostructural analysis of ITO thin film.

RESULT AND DISCUSSION

Work Function

In any OLED, injection of charges from cathode and anode into the emissive organic layer depends on the matching of energy levels. So cathode and anode work function has to match with highest occupied molecular orbit (HOMO) for hole injection and lowest unoccupied molecular orbit (LUMO) of its organic material for electron injection [13]. Practically, proper material combination has to be found for both cathode and anode transport layer. Energy level diagram for polymer OLED devices is shown in figure 1.

ITO is used as anode, the work function of ITO has to match with HOMO level of hole transport layer (PEDOT: PSS for polymer devices) which has work function range (2.4ev-5.0ev) The measured work function of ITO film deposited on glass and plastic substrate with RF magnetron sputtering comes out to be ~4.6ev which is comparable with commercially available ITO on glass having work function ~4.5ev. Plasma treatment of ITO coated on plastic substrate increases the work function to 4.7ev [14]. From the energy level diagram it is clear that these ITO films' work function is good enough for the fabrication of FOLED with polymers as well as small molecules.

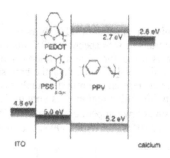

Figure1. Energy level diagram of Polymer OLED [13]

Sheet Resistance

Sheet resistance of ITO coated films on plastic substrates is listed in Table1. Some values of sheet resistance on glass and polycarbonate substrate are also mentioned. Temperature and gas pressure were held constant during these depositions. Sheet resistance of ITO film decreases with deposition time as the thickness of the film increases. We found that sheet resistance becomes half when the thickness of the film is approximately doubled indicating that thickness of the film is directly proportional to the deposition time. Variation in resistivity is from 2.4×10^{-4} to $2.7 \times 10^{-4} \Omega$ cm. This is due to the fact that as thickness increases we get more uniform ITO film and also the roughness due to substrate has less impact. We get sheet resistance only due to ITO film. The change in sheet resistance after one month is also shown in table1. We saw that resistance increases with time. This is due to the cracks formed in the substrate, which increases the resistance. It is clear from the table that increase in sheet resistance of thinner film is larger compared to the thicker ITO film [15]. There is an increase in sheet resistance of ITO film but after certain time (~one month) its value get saturated. The values of sheet resistance after one month shows that still we can use these substrates for FOLED fabrication purpose because some studies have shown sheet resistance in this range [16].

No.	Deposition Time (min.)	Sheet resistance as deposited ($\Omega/\square$)	Thickness (Å)	Resistivity (Ω-cm)	Sheet resistance ($\Omega/\square$) After one month
1.	10.0	14.0	1745.0	2.4×10^{-4}	55.5
2.	15.0	10.5	2523.0	2.6×10^{-4}	-
3.	20.0	8.8	3139.0	2.7×10^{-4}	31.3
4.	30.0	6.8	4026.0	2.7×10^{-4}	25.0
5.	20.0	8.2 (on glass)			
6.	20.0	33.3 (Polycarbonate)			

Table1. Sheet resistance, thickness and resistivity of ITO film at 80^0 C temperature for different deposition time.

Optical Transmittance

Figure 2 is the transmission spectra of ITO film on plastic deposited at different temperature. Figure 3 shows the transmission spectra of ITO film for different deposition time at 80°C. A transmission spectrum is taken for 300nm -1000nm range. Transmission data in the figure 2 & 3 shows the transmission of ITO including the bare substrate. The transmission of plastic substrate is also given for the reference. From the graph it is clear that Mylar plastic sheet has only 85% of transmission in the visible region. It is very difficult to get ITO film with 95% transmittance in the visible region at low temperature and resistivity of 10^{-4} Ω cm. [17]. Here we get stable resistivity of the order of 10^{-4} Ω cm and 87-90 % transmission from ITO thin film.

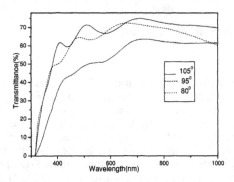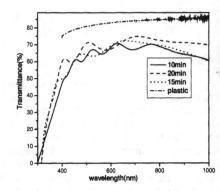

Figure 2. Transmission Spectra of ITO on plastic at different Temperature.

Figure 3. Transmission Spectra of ITO film on plastic at 80° C for different deposition time.

It is observed that transmission decreases very fast near ultraviolet region (300nm-400nm) which is because the band gap of 3.6-3.85ev in In_2O_3 [18,19]. Low transmission of ITO film on plastic substrate is due to the defect states in the band gap region. This suggested that transmission could be improved by annealing at low temperature but for a longer time as reported earlier [17].

X-Ray Diffraction Results

X ray diffraction results of ITO on plastic as well as glass substrates are shown in figure 4. An XRD spectrum of plastic (figure.4a) has a [111] peak at 26.09°C. ITO on plastic substrates shows a prominent peak at around 26.09 (25.83 at 80° C & 26.13 at 105°C) is due to plastic figure(4c &4d). It is clear that as the deposition temperature increases a [111] peak of lower intensity is observed at the point where the peak of plastic appears and also a [400] peak at 35.02 is also observed. So we can say that we get an amorphous but highly ordered film of ITO at lower temperature on plastic substrate as reported earlier [20]. As temperature increases some crstanillity is observed in the film. In case of ITO film on glass substrate there are peaks as [222] at 30.43, [400] peak at 34.00 [4 3 0] at 45.122 showing crystal nature of the ITO film (figure 4b).

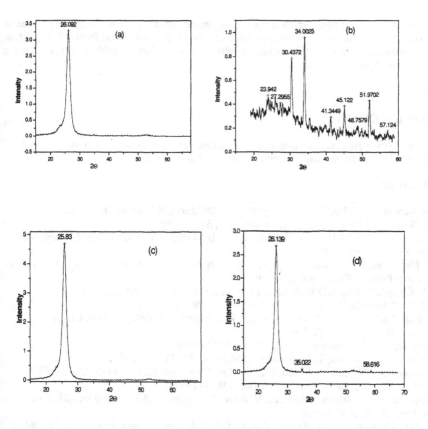

Figure4. XRD spectrum of (a) Plastic substrate (b) ITO on glass at 80^0 C (c) ITO on plastic substrate at 80^0 C (d) ITO on plastic substrate at 100^0 C.

CONCLUSIONS

For flexible organic light emitting displays, uniform sheet resistance, high transmission and low resistivity of ITO thin film is required. Here, ITO thin films on plastic were prepared with RF magnetron sputtering at low power and low temperature. From the experimental results based on work function, sheet resistance, resistivity, optical transmission and XRD of ITO on plastic substrate we found that poor transmission in visible region is due to surface roughness of ITO. From XRD results we see that substrate has a large influence on the properties of ITO thin film. Stable and low resistivity (~2.4x10^{-4} Ω-cm) is comparable with the values reported by Lee [21]. Films at low temperature are amorphous in nature. But at high temperature we get some

143

crystalline nature of the film, which causes large amount of defects and charge traps. Due to which we get less transmission. From all above results we concluded that from RF magnetron sputtering one could make ITO film on plastic substrates, which have suitable properties to make flexible organic light emitting devices.

ACKNOWLEDGMENTS

This work is supported by Department of Science and Technology, Govt. of India and done in SAMTEL center for display technologies, IIT Kanpur India.

REFERENCES

1. J.H.Burroughes, D.D.C.Bradley, A.R.Brown, R.N.Marks, K.Mackay, R. H.Friend, P.L.Burns, A.B.Holmes, *Nature (London)*, **347**, 539 (1990).
2. H. Kim, A. Pique, J.S.Horwitz, H. Mattoussi, Z.H.Murata, Z.H. Kafafi. D.B. Chrisey, *Appl. Phys. Lett.*, **78**, 1050 (2001).
3. H.L. Hartnagel, A.L. Dawer, A.K. Jain, C. Jagdish, *Semi conducting Transparent Thin Films, Institute of Physics, Philadelphia,* **91** (1995).
4. K.L. Chopra, S. Major, D.K. Pandya, *Thin Solid Films*, **102**, 1 (1983).
5. S. Seki, Y Sawada, T. Nishide, *Thin Solid Films*, **388**, 22 (2001).
6. H. Kim, J.S.Horwitz, G. Kushto, A. Pique, Z.H. Kafafi, C.M.Gilmore, D.B. Chrisey, *J. Appl. Phys.*, **88**, 6021 (2000).
7. Y. Hoshi, R. Ohki, *Electrochimica Acta*, **44**, 3927 (1999).
8. Y. Shigesato, S. Takaki, T. Haranoh, *J. Appl. Phys.*, **71** (7), 3356 (1992).
9. S.B. Lee, J.C. Pincenti, A. Cocco, D.L. Naylor, *J. Vac. Sci. Tech.*, **A11** (5), 2742 (1993).
10. Y. Shigesato, D.C.Paine, *Thin Solid Films*, **238**, 44 (1994).
11. T.Oyama, N. Hashimoto, J. Shimizu, Y. Akao, H. Kojima, K.Aikawa and Suzuki, *J. Vac. Sci. Tech. A*, **10**, 1682 (1992).
12. H. Kim, J.S. Horwitz, G.P. Kushto, Z.H. Kafafi, D.B. Chrisey, *Appl. Phys. Lett.*, **79**, 284 (2001).
13. R. H. Friend, *Pure Appl. Chem.*, **73**, 425 (2001).
14. Furong Zhu, *J. of SID*, **11**, 605 (2003).
15. D.R. Cairns, R.P.W. II, D.K.Sparacin, S.M.Sachsman, D.C.Paine, G.P. Crawford, R.R.Newton, *Appl. Phys. Lett.*, **76**, 1425 (2000).
16. Suchandra Bhaumik, A. K. Barua, *Jpn. J. Appl. Phys*, **42**, 3619 (2003).
17. T. Minami, H. Sonohara, T. Kakumu, S. Takata, *Thin Solid Films*, **270**, 37 (1995).
18. A.K. Kulkarni, S.A. Knickerbocker, *J. Vac. Sci. Tech. A*, **14**, 1706 (1996).
19. M. Buchanan, J.B. Webb, D.F. Williams, *Appl. Phys. Lett.*, **37**, 213 (1980).
20. V. Craciun, D. Craciun, X. Wang, T.J. Anderson, R.K. Singh, *Thin Solid Films*, **453**, 256 (2004).
21. H. Lee, I. G. Kim, S.W. Cho, S.H.Lee, *Thin Solid Films*, **302**, 25 (1997).

Mater. Res. Soc. Symp. Proc. Vol. 869 © 2005 Materials Research Society   D2.1

Study of sputtered Hafnium oxide Films for Sensor Applications

H. Grüger, C. Kunath, E. Kurth, S. Sorge, W. Pufe
Fraunhofer IPMS, Grenzstr. 28, 01109 Dresden, Germany

Abstract

In this paper results of the deposition and annealing of hafnium oxide thin films are reported. Due to the sensor application in mind, thicknesses between 30 and 150 nm have been deposited by r.f. sputtering of a high purity oxide target. Annealing has an important influence on the layer structure, stress and application correlated properties. A detailed understanding of the layer preparation is necessary to adjust deposition and annealing. After deposition the layers are predominately amorphous, annealing leads to textured layers with monocline or orthorhombic phases.
Besides gas sensor applications optimized layers may serve as protective coating or combined with a second material to multi layer stacks as high reflective dielectric mirror.

Introduction

Transition metal oxides have gathered importance for different applications. Some examples are surface coatings, dielectric layers for semiconductors or as sensitive material for thin film sensors [1, 2].
Important for these applications are the manifold exceptional material properties of these oxides like high ε, good temperature stability or mechanical and chemical inertness. Due to the wide range of applications, a broad variety of thickness has to be deposited. Atomic layer deposition (ALD) [3] and metal organic chemical vapor deposition (MOCVD) [4] has been used for thin layers. Physical vapor deposition (PVD) like sputtering is suitable for thicker layers.
Today the main focus are ultra thin layers for semiconductor devices [5] but nevertheless sensor applications use the same material for its favorable electrochemical properties.

Experimental

Hafnium oxide films with a thickness range between 30 and 150 nm have been deposited by r.f. sputtering from a high purity oxide target in Argon / Oxygen mixtures [6, 7]. 150mm wafers served as substrate. During deposition no additional substrate heating was applied. Thicker layer were deposited continuously or in multi step deposition without interruption of the vacuum.
After deposition thermal treatment up to 1000°C was possible either by RTA with 50 K/s heating rate or in a standard CMOS oven process with 3K/min. The total annealing time was set to 30 min typically 10 min in Oxygen and 20 min in Nitrogen or Argon.
After finalization of the process the stress was measured by bow measurements. Surface analysis was applied by AFM and critical dimension SEM. For further characterization wafers were cut into pieces. X-ray diffraction and TEM have been used to determine crystalline fraction, structure and orientation. The chemical and electrochemical characterization was done on samples in the laboratory.

Layer properties

From earlier work [6, 7] it is well known, that the layer properties are strongly influenced by deposition and annealing parameters. First of all, the layer stress has been evaluated. After deposition typically compressive stress is observed, the thicker the layers the higher the stress. Multi deposition of thin layers leads to a reduction of stress (fig. 1).

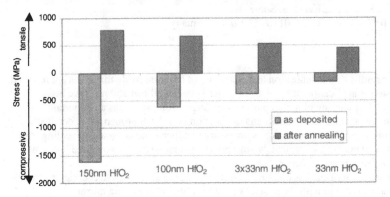

Fig. 1: Layer stress of HfO_2 layer with different thickness after deposition and after annealing

After annealing typically tensile stress is observed with a similar tendency of the thickness. Additional experiments with the same thickness of 100nm deposited either in one or in three steps and different annealing procedures reveal, that the stress can be reduced significantly by a proper selection of the parameters. The behavior is different from that of Ta_2O_5 examined earlier [6]. The layer stress is important as it improves film adhesion and minimizes the probability of cracking.

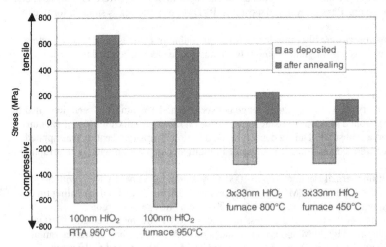

Fig. 2: Layer stress of 100nm HfO_2 prepared and annealed differently

To evaluate the layer structure, high resolution TEM has been used. Figure 3a shows a cross sectional view through a 150 nm HfO₂ layer as deposited on an oxidized silicon substrate. Close to the surface the layer shows a crystalline structure, close to the substrate the layers are predominately amorphous.

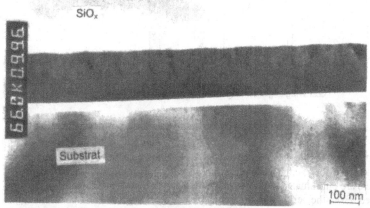

Fig. 3a: TEM cross section through a 150nm HfO₂ layer as deposited on oxidized silicon

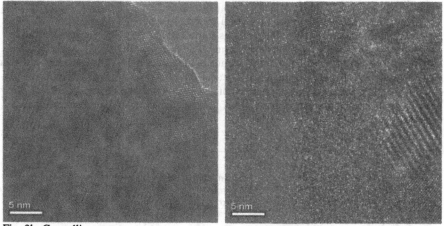

Fig. 3b: Crystalline part   Fig. 3c : Transition area

Figure 3b shows an high resolution image of the crystalline part, bright spots indicated lines of Hafnium atoms. Figure 3b give an image of the transition area with the beginning crystallization in a high magnification.
X-Ray diffraction pattern have been collected with 0.5 and 1° incidence angle to gain information of crystal fraction in different depths. The information depth was 2..70nm for 0.5° and 80..140nm for 1° incidence angle.
Measured XRD pattern have been compared with powder diffraction files from a data base (fig. 4).

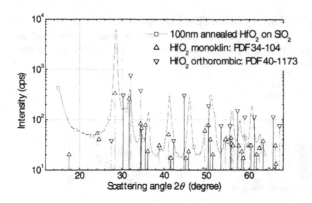

Fig. 4: Measured pattern of an annealed HfO$_2$ layer and data from powder-diffraction-files

A difference of two up to three degrees between the peak positions from the files and the measured peaks were observed. Two layer properties may impair the resolution of sharp lines: First there are only small crystallites embedded in amorphous matrix causing the broad peaks measured. From the peak width the grain size has been estimated to 15.5..18nm after deposition and 21..25nm after annealing. Second the grain orientations are not uniformly distributed compared to the powder references. The layers have always been found textured. Only the assumption of the occurrence of both the monocline and orthorhombic crystal phase explains the measured peaks. A possible third crystal phase of HfO$_2$ is cubic. This modification forms preferably at temperatures above 1500°C. At lower temperatures the cubic phase can be stabilized by impurity atoms as discussed elsewhere [8]. Also certain sputtering conditions may support the formation of the cubic HfO$_2$ [8]. In the layers described here, the cubic phase could not be proven, but may also be present in low percent region. Significant differences between the amount of amorphous and crystalline material result from annealing. In figure 5 the pattern from a layer as deposited and a layer after annealing are compared.

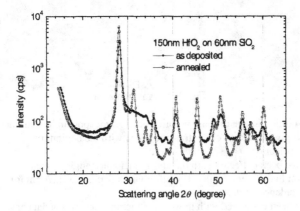

Fig. 5: Comparison of as deposited and annealed layers (0..70nm depth)

The relation between the amorphous phase and the crystalline phase has been estimated from the modulation of the spectra between peak base and peak top. The relation has been found less dependent from the silicon or silicon dioxide substrate. After deposition the HfO_2 layer has always been found amorphous on the substrate. The initial layer thickness rules the grain growth much more. As pointed out in the TEM analysis HfO_2 spontaneously organizes into a crystalline phase in layers thicker than 50nm. With increasing depth the amount of amorphous material increases (see table 1 for details). This result is in good accordance with the TEM-analysis.

Table 1: Estimated fractions of amorphous and crystalline HfO_2

HfO_2 layer conditions	0,5° incidence 0..70nm depth	1° incidence 80..140nm depth
100nm on Si as deposited	30	53
150nm on Si as deposited	20	-
3x50nm on SiO_2 as deposited	15	32
150nm on SiO_2 as deposited	10	17
50nm on SiO_2, annealed	5	9
100nm on SiO_2, annealed	5	7

Annealing increases the crystallization of the layer until the layers appear thoroughly crystallized in TEM images. Further, annealing changes the ration between monocline to orthorhombic phases.
The monocline phase features strains with $3..5*10^{-3}$ in non annealed layers and $-7..-8*10^{-3}$ in annealed layers. Positive numbers indicate larger atom distances, negative numbers shorter distances. This result is in accordance with the stress results, where pressure stress in layers as deposited change into tensile stress after annealing. Strain increases $4*10^{-3}$ with depth, indicating increasing atom distances as the material becomes more and more amorphous.
Results of the sputter parameter variation on the layer properties are summarized in table 2.

Table 2: Summary of layer properties for different deposition and annealing parameters

Deposition parameter / Layer property	Deposition rate	Homo- genity	Grain size	Stochio- metry	Stress
Higher sputter power			+	-	
• Low pressure / low oxygen	+	+			compressive -
• High pressure / low oxygen	o	+			o
• Low pressure / high oxygen	+	+			compressive+
• High pressure / high oxygen	+	-			compressive+
Higher sputter pressure	-	+	-	+	compressive -
More Oxygen	-	-	-	+	compressive -
Higher target to substrate distance	-	+	-	o	compressive +
Higher annealing temperature			+	+	tensile +
Longer annealing time			+	+	tensile +

+: increase, -: decrease, o: constant

Summary and discussion

Deposition and annealing of Hafnium oxide layers in a thickness range between 30 and 150nm has been evaluated with structural, mechanical and chemical properties in mind. The layer stress has been found to be strongly dependant on the deposition and annealing. Thicker layer produce higher stress. Multi step deposition of several thin layers reduces the stress . Reduced annealing temperatures (950 to 450°C) also leads to lower stress. Structural analysis shows, that annealing starts a complete crystallization of the former predominately amorphous layers. Two crystal phases occur together: monocline and orthorhombic is detected by XRD. The cubic phase, typical for very high temperatures, that might indicate impurities or special non equilibrium effect during deposition has not been detected. This can be due to the high purity of the deposition. Compared to Ta_2O_5, the deposition and annealing of HfO_2 shows some important differences, nevertheless both materials serve quite similar applications.

Acknowledgements

The authors would like to thank the clean room staff for preparation work and Dr. Frank Eichhorn from FZ Rossendorf for XRD analysis.

References

1. Yu Lebedinskii et al., "Silicide formation at HfO_2/Si and ZrO_2/Si interface induced by Ar^+ ion bombardment" *Mat. Res. Soc. Symp. Proc.* 768 (2004) E3.25, 165

2. S. Duenas et al., "On the interface quality of MIS structures fabricated from Atomic Layer Depostion of HfO_2, Ta_2O_5 and Nb_2O_5-Ta_2O_5-Nb_2O_5 dielectric thin films" *Mat. Res. Soc. Symp. Proc.* 786 (2004) E3.18 1

3. Kaupo Kukli et al. "Effect of selected atomic layer deposition parameters on the structure and properties of hafnium oxide films" *J. Appl. Phys.* 96 (2004) 5298

4. M. Balasubramanian et al. "Wet etching characteristics and surface morphology of MOCVD grown HfO_2 films" *Thin Solid Films* 462-463 (2004) 101

5. K. Onishi, C. Kang, R. Choi, H.-J. Cho, Y. Kim, S. Krishnan, J. C. Lee "Performance of polysilicon gate HfO_2 MOSFETs on (100) and (111) Silicon substrates" *IEEE Electron Device Letter* 24 (2003) 254

6. H. Grüger, C. Kunath, E. Kurth, W. Pufe, S. Sorge "Improved structural properties of sputtered hafnium dioxide on silicon and silicon oxide for semiconductor and sensor applications" *Proceedings of the Materials Research Society* Fall Meeting 2003

7. H. Grüger, C. Kunath, E. Kurth, S. Sorge, T. Pechstein "High quality r.f. sputtered metal oxides (Ta_2O_5, HfO_2) and their properties after annealing", *Solid Films* 447-448 (2004) 509-515

8. R. Manory, T. Mor, I. Shimizu, S. Miyake, G. Kimmel "Growth and structure control of HfO_{2-x} films with cubic and tetragonal structures obtained by ion beam assisted deposition" *J. Vac. Scie. Technol.* A20 (2002) 549

Mater. Res. Soc. Symp. Proc. Vol. 869 © 2005 Materials Research Society   D2.7

Co-firing of Low- and Middle Permittivity Dielectric Tapes in Low-Temperature Co-fired Ceramics

Jae-Hwan Park, Young-Jin Choi, and Jae-Gwan Park
Materials Science and Technology Division, Korea Institute of Science and Technology,
PO Box 131, Cheongryang, Seoul 130-650, Korea

Abstract

Compatibility in cofiring low-κ and middle-κ based on a same glass composition for the hybrid LTCC material systems was investigated. Designing of appropriate lithium borosilicate glass frit systems allowed us to develop an optimized glass frit system used for low-κ and middle-κ dielectric fillers, commonly. The effects of glass addition on the densification and electrical properties in low-κ and middle-κ dielectric materials were also examined. Controlling glass compositions and contents we tried to match low-κ and middle-κ tapes physically and chemically.

Introduction

As the next information society approaches, it accompanies a new progressing communication system technologies, development of wireless, digitalized, high technological, and portable electronic devices and requires the next generation of following chip-device technology with miniaturization and multi-functional integration. [1-2]. LTCC multilayer structures have been generally used as a concept of a 3D wiring circuit board to date, using low-permittivity dielectric compositions (typical permittivity is 4~9). To realize highly integrated and multi-functional LTCC modules, however, it is important to incorporate passive components such as resistors, capacitors, inductors, and other functional parts in LTCC multilayer structure. Eventually, LTCC packages will proceed to an all-in-one type module, which incorporate various passive components and functions, as illustrated in Fig.1.

Among various components which could be realized in LTCC packages, resonators and internal capacitors are the most desirable. The internal capacitors are required typically to realize circuit decoupling monolithically in LTCC packages. The typical forms of the resonators are a strip or a microstrip line of quarter wavelength on the substrate. The appropriate permittivity range for the resonators and the internal capacitors is 20~500.

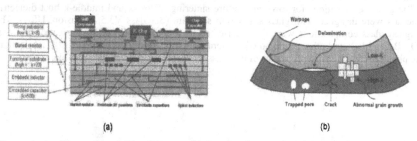

(a)   (b)

Figure 1. An illustration of (a) a future multi-functional LTCC module and (b) defects induced by the physical and chemical incompatibilities between hetero LTCC layers.

There have been several investigations on the middle-κ LTCC compositions [3-4]. Some microwave dielectric materials were reported to be sintered at the lower temperature range with addition of an appropriate amount of sintering aids such as oxides or glass frit. As illustrated in Fig. 1 (a), the low-κ as well as the middle-κ materials is required for realizing highly complicated and multifunctional modules. When the low-κ and middle-κ LTCC tapes are laminated and cofired in a module, the physical and chemical compatibilities between low-κ and middle-κ hetero layers should be considered carefully in the development of middle-κ LTCC materials. As shown in Fig. 1(b), there might be delaminations and/or warpages between hetero layers. When the chemical reactions between the glass frit of low-κ and middle-κ layers are limited, and/or there is a large difference between relative shrinkages of hetero layers, complete delaminations would occur. When there are some mismatches in final relative shrinkage and shrinkage profile (i.e. characteristic curve of shrinkage as a function of temperature), warpages may occur. Furthermore, there might be various defects such as trapped pores, cracks, and abnormal grain growth when there are over-firing, thermal shrinkage mismatch, and other chemical/physical incompatibilities.

We already developed several middle-κ LTCC material systems. However, unfortunately, we found that those compositions are not matched with the low-κ substrate tapes physically and/or chemically. In some cases, complete delaminations would occur between hetero-layers. Warpages, due to the densification profile mismatch, frequently happens. Thus, we concluded that it is very important to consider low-κ and middle-κ LTCC compositions simultaneously considering their sintering properties and chemical properties.

As for low-κ LTCC tapes, B_2O_3-SiO_2 based glass systems were typically used whereas the glass frit with high Li_2O content (typically, Li_2O >30 wt%) were used for the fabrication of middle-κ LTCC tapes. However, in this work, we designed low-κ and middle-κ dielectric composition based on a same glass composition so that the physical and chemical compatibilities between low-κ and middle-κ hetero layers could be obtained. By designing a lithium borosilicate glass frit system carefully, we developed optimum glass frit system for both low-κ and middle-κ dielectric fillers. The effects of glass addition on the densification, electrical properties, and other issues in low-κ and middle-κ dielectric materials were examined.

Experimental details

We selected two host dielectric materials for low-κ and middle-κ LTCC compositions. Mg_2SiO_4, abbreviated M2S hereafter, which exhibits low permittivity and dielectric loss, was selected for low-κ composition [5]. $BaTi_4O_9$-based dielectrics, hereafter BT4, have been proved as good middle-permittivity materials for LTCC system [6].

The detailed compositions and their sintering and dielectric properties, which are measured from our preliminary experiments, are summarized in table 1. The M2S was fabricated by a solid state sintering process with the oxide forms of reagents (MgO and SiO_2, 99.9% purity). In our preliminary experiments, all the dielectric compositions could be densified at higher than 1350°C and showed the relative density of more than 98% and good dielectric properties, without glass addition, as shown in table 1.

The glass compositions for low-temperature sintering of low-κ and middle-κ host dielectric materials were designed with glass simulation software (SciGlass V3.5, Scivision, USA). The designed glass compositions were then fabricated by a conventional glass fabrication process. SiO_2, B_2O_3, Li_2O, and other oxide powders were used as starting materials for glass fabrication. The physical and electrical properties of the glass frit were summarized also in table 1.

(a)

Host dielectric materials

		Sintering properties			Dielectric properties			
Abbreviation	Material	Optimum sintering temp. [°C]	Relative density [%]	Linear shrinkage [%]	Permittivity [1]	tan d [%] [1]	Q·f [GHz] [1]	τ_f [ppm/°C] [1]
M2S	Mg₂SiO	1450	97.3	16.8	7.0	<0.02	43000	−60
BT4	BaTi₄O₉	1350	98.7	16.9	38.6	<0.05	44500	+3

[1] Commercial powder (MWF-38, Hayashi Chemical Industry, Co. Ltd, Japan)
[2] Dielectric properties were measured using a network analyzer in the frequency range of 9~16 GHz.
[3] Tan δ [%] were measured using an impedance/gain-phase analyzer (1 MHz).

(b)

HK series glass frits

Physical properties	Density [g/cm³] : 2.31~2.52
	α_{CT} [10⁻⁶/°C] : 6.3~12.5
	Glass transition point, T_g [°C] : 371~575
	Glass softening point, T_s [°C] : 390~627
Dielectric properties	Permittivity [1] : 7.0~9.0
	tan d [%] [1] : 0.2~0.5
	Q·f [GHz] [1] : 1200~2300
	τ_f [ppm/°C] [1] : −83~−223

[1] Dielectric properties were measured using a network analyzer in the frequency range of 12~16 GHz.
[2] Tan δ [%] were measured using an impedance/gain-phase analyzer (1 MHz).

Table 1. (a) The sintering and electrical properties of the host dielectric materials used in this study. (b) The physical and electrical properties of the glass additives used in this study.

The dielectric compositions and 5~30 wt% of glass frit were mixed by ball-milling for 24 h. The slurry was then dried at 90°C for several hours and granulated with 3 wt% of PVA (poly-vinyl alcohol) solution. The granulated powders were pressed into disk shapes and sintered in the temperature range of 800~950°C for the evaluation of densification. Dielectric constant, microwave quality factor, and temperature coefficient of resonant frequency (τ_f) were measured by a parallel plate resonator method and a cavity resonator method with a network analyzer (8720C, HP, USA) [7]. In some cases, the dielectric properties were measured by an impedance analyzer (4194, HP, USA) at 1MHz. For the evaluation of the properties of thick films, the tapes were prepared by the mixing of host dielectrics, glass frit, and vehicle system (Ferro, #73225).

Results

The design of glass frit is important to lower the densification temperature of the host dielectric materials to less than 875°C still maintaining acceptable dielectric properties. As for low-κ LTCC substrate, in general, 30~60 wt% of high borosilicate systems (SiO₂+B₂O₃ >80 wt%) were mixed with fillers. As for middle-κ LTCC compositions, the glass frit with high Li₂O content (typically, Li₂O >30 wt%) were used. Basically, therefore, there should be chemical incompatibilities between low-κ and middle-κ compositions due to such differences in glass composition. Our strategy is to apply the equal amount of glass frit (<20 wt%), based on Li₂O-B₂O₃-SiO₂, to both the low-κ and middle-κ dielectric fillers so that the physical and chemical compatibilities between low-κ and middle-κ hetero layers could be obtained.

To realize acceptable densification and dielectric properties at less than 875°C, it is important to design the glass compositions which have lower glass transition temperature (T_g) and dielectric loss. Throughout our repeated simulations and experiments, it was concluded that the glass compositions should have T_g of less than 500°C and dielectric loss of less than 1% to guarantee the temperature of densification less than 875°C still maintaining acceptable dielectric properties. Among those tested, we selected a series of glass frit as listed in table 1 for this study.

As the content of Li₂O and B₂O₃ increase, the T_g of the frit decreases whereas the dielectric loss increases. Thus, the glass frit with higher Li₂O and B₂O₃ were appropriate for the sintering aids for the middle-κ host dielectric material. The glass compositions with higher SiO₂ content were appropriate for low-κ dielectric composition, i.e. M2S.

As the low-κ LTCC substrates have been used for 3-D routing substrates, the lower permittivity is desirable for minimizing signal delay and the lower dielectric loss is preferred to guarantee higher Q factors in the circuits. Typically, the low-κ LTCC materials are composed with fillers (e.g. alumina or silica) and high borosilicate glass. However, in this study, we have searched low-κ fillers which is appropriate for applying the high lithium borosilicate glass frit (Li$_2$O > 20 wt%). Eventually, we could find M2S is more appropriate material than typical alumina for applying Li$_2$O-B$_2$O$_3$-SiO$_2$ glass system. The detailed compositions and their sintering and dielectric properties are summarized in table 1 (a).

Glass contents [wt%]	Firing temp. [°C]	Dielectric properties			
		Permittivity [†]	tan δ [%] [‡]	Q×f [GHz] [†]	τ$_f$ [ppm/°C] [†]
5	900	6.8	0.09	16000	-60
7	875	6.9	0.14	14300	-69
10	875	7.0	0.24	13300	-77

Table 2. A summarized table of the sintering behaviors and electrical properties of the M2S ceramics with glass addition.

Figure 2 shows the effects of glass addition on the low-temperatire sintering behaviors in M2S ceramics. As the amount of the glass frit increases, both the linear shrinkage and bulk density increases at the temperature range of 825~875°C as shown in Fig. 2 (a)-(b). When 7 wt% of the frit was added, significant densification was achieved at 875°C. As the amount of added frit increases, the higher densification was achieved at less than 850°C whereas a slight decrease of shrinkage was observed at the temperatures higher than 900°C. When 7 wt% of glass frit was added at 875°C, the density of 3.02 g/cm^3 (relative density 97%) and linear shrinkage 16% were obtained. A scanning electron microscope (SEM) image shows that a sufficient densification was obtained without pores. The details of electrical properties are shown in table 2. When 7 wt% of the frit was added at the firing temperature of 875°C, the permittivity of 6.9 and Q of 700 were obtained.

For the development of the middle-κ LTCC material systems, it is required that the materials should be densified at less than 875°C still maintaining high permittivity. In this study, we tested several microwave dielectric materials which have permittivity of 20~100 for the middle-κ LTCC material systems. We found that the BT4 could be appropriate for low-temperature sintering with small amount of glass frit. The detailed compositions and their sintering and dielectric properties are summarized also in table 1 (a).

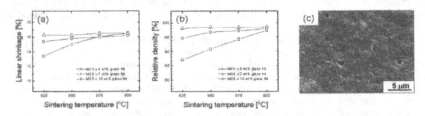

Figure 2. The effects of glass content and sintering temperature on the sintering behaviors in M2S ceramics. (a) Linear shrinkage, (b) relative density, (c) a SEM image of a sample sintered at 875°C with 7 wt% of glass frit.

The effects of glass addition on the low-temperature sintering of BT4 were examined. The sintering behaviors of BT4 with addition of glass frit are shown in Fig. 4. As the amount of the glass frit increases, both the linear shrinkage and bulk density increases at the temperature range of 825~850°C as shown in Fig. 3 (a)-(b). When 15 wt% of the frit was added, significant densification was achieved at 875°C. As the amount of added frit increases, the higher densification was achieved at less than 875°C whereas the final density of the sintered body decreases at the temperatures higher than 875°C. When 15 wt% of glass frit was added at 875°C, the density of 4.11 g/cm^3 (relative density 98%), linear shrinkage 16%, open porosity 0.2% were obtained. The fact that the density decreases with increasing the amount of frit in the temperature range of 875~900°C could be ascribed to the low density of glass itself. A SEM image shows that a sufficient densification was obtained without pores.

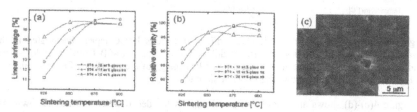

Figure 3. The effects of glass content and sintering temperature on the sintering behaviors in BT4-based ceramics. (a) Linear shrinkage, (b) relative density, and (c) A SEM image of a sample sintered at 875°C with 10 wt% of glass frit.

For the evaluation of the physical and chemical matching between hetero- layers, low-κ and middle-κ LTCC tapes were prepared. Table 3 shows selected compositions of low-κ and middle-κ LTCC tapes with different content and composition of glass. All the physical and electrical properties were evaluated from the sintered tapes of each composition. Low-κ tapes exhibits the permittivity of 5.6~6.6 and Q value of 2000~4000. It is noteworthy that the Q value in this study is far higher than those of the tapes commercially available. The permittivities of 22~29 were measured in middle-κ tapes.

Code	Ceramics composition		Glass frit		Relative density [%]	Permittivity (k) [1 MHz]	tan δ (%) [1 MHz]
	Type	Contents [wt%]	Type	Contents [wt%]			
L03	M2S	95	HK6	5	95.4	5.59	0.047
L04	M2S	90	HK6	10	98.8	6.63	0.045
L07	M2S	90	HK7	10	99.4	6.60	0.028
M04	BT4	90	HK8	10	95.8	28.7	0.180
M05	BT4	85	HK8	15	96.3	24.2	0.450
M09	BT4	85	HK9	15	97.8	22.9	0.520

Table 3. A summarized table of the sintering behaviors and the electrical properties of the selected LTCC tapes.

Figure 4(a)-(b) shows the selected dilatometric shrinkage profiles of the low-κ and middle-κ tapes. As the added amount of glass frit increases, the on-set temperature of densification decreases downward. The content of Li$_2$O and B$_2$O$_3$ in the glass composition increases, the shrinkage profile also decreases. Thus, we can get similar profile of densification at a combination of L04-M05, by adjusting glass composition and amount of added glass. As shown in the figure, the on-set temperature and densification profile of the L04-M05 combination were most matched.

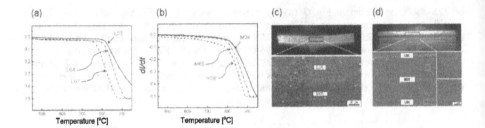

(a)   (b)   (c)   (d)

Temperature [°C]   Temperature [°C]

Figure 4. (a)-(b) Dilatometric shrinkage profiles of various LTCC tapes investigated in this study; (a) Low-κ LTCC tapes, and (b) Middle-κ LTCC tapes. (c)-(d) Cofiring of low-κ and middle-κ LTCC types. An optical image and a SEM image of the interface of (c) the bilayered A-B structure and (d) the sandwitched A-B-A structure.

Figure 4(c)-(d) shows an example of matching. As the densification profiles of low-κ and middle-κ LTCC tapes were matched well as shown in Fig. 4(a)-(b), the warpages and delamination were minimized. Furthermore, as we applied the common glass frit, based on $Li_2O-B_2O_3-SiO_2$, there were no delaminations observed. Fig. 4(c) shows a case of A-B structure whereas (d) represents a case of A-B-A sandwitched structure, where the minimal warpage is observed. As a common glass system was used in this work, it is observed that there are no cross- diffusions in the interface as shown in Fig. 4(c) ~ (d).

Summary

In the development of low-κ and middle-κ hybrid LTCC material systems, physical and chemical compatibilities between hetero-layer are very important. In this regard, we designed low-κ and middle-κ dielectric composition based on a same glass composition so that the compatibilities between low-κ and middle-κ hetero layers could be obtained. By designing a lithium borosilicate glass frit system carefully, we developed optimum glass frit system for both low-κ and middle-κ dielectric fillers. The low-κ and middle-κ LTCC tapes, based on a same glass system, exhibit excellent compatibilities in physical and chemical matching.

References

1. R. R. Tummala, *J. Am. Ceram. Soc.*, **74**, 895 (1991).
2. J. Müller, H. Thust, and K. H. Drüe, *International Journal of Microcircuits and Electronic Packaging*, **18**, 200 (1995).
3. H. Jantunen, R. Rautioaho, A. Uusimaki, and S. Leppavuori, *J. Am. Ceram. Soc.*, **83**, 2855 (2000).
4. T. Takada, S. F. Wang, S. Yoshikawa, S.-J. Jang, and R. E. Newnham, *J. Am. Ceram Soc.*, **77**, 2485 (1994).
5. D. J. Masse, R. A. Purcel, D. W. Readey, E. A. Maguire, and C. P. Hartwig, *Proc. IEEE*, **59**, 1628 (2001).
6. C. F. Yang, *Jpn. J. Appl. Phys.*, **38**, 3576 (2001).
7. B. W. Hakki and P. D. Coleman, *IRE Microwave Theor. Tech.*, **MTT-8**, 402 (1991).

Mater. Res. Soc. Symp. Proc. Vol. 869 © 2005 Materials Research Society

Application of Magnetic Ferrite Electrodeposition and Copper Chemical Mechanical Planarization for On-Chip Analog Circuitry

Cody Washburn[1], Daniel Brown[2], Jay Cabacungan[1], Jayanti Venkataraman[2] and Santosh K. Kurinec[1]
[1]Microelectronic Engineering, [2]Electrical Engineering
Rochester Institute of Technology, Rochester, NY 14623

ABSTRACT

Inductors are important components of analog circuit designs, from matching circuitry to passive filters. In this study, the application of electrophoretically deposited nano-ferrite material has been investigated as a technique to increase the inductance of integrated copper planar inductors fabricated using copper plating and chemical mechanical planarization. Sintered Mn-Zn ferrite particles are suspended in a medium of isopropyl alcohol with magnesium nitrate and lanthanum nitrate salts. The transportation of the particles to the substrate surface is assisted by applied electric field and particles adhere to the substrate surface by a glycerol based surfactant. Electrophorectic deposition process forms a self aligned polymeric thin film on the surface of a p-type silicon substrate selectively with respect to copper. This ferrite deposition method yields high selectivity to the inductor coils and patterned silicon substrates compatible with standard silicon technology.

INTRODUCTION

Monolithic integration of passive circuit components such as inductors, capacitors and resistors can provide on-chip circuitry capable of performing a number of analog signal processing functions. When combined with CMOS technology, the integration of passives will result in greater device functionality along with reduced package size. The optimization of on chip passive components is critical to improve performance of silicon based analog technology. Planar inductor coils are a common configuration for incorporating highly inductive elements in silicon based technology. The planar structure is particularly easy to incorporate in a CMOS process, by utilizing the top level of metal interconnect. For optimal performance, copper is used to define the inductor pattern in order to minimize resistance. The mutual coupling of the magnetic field between adjacent traces of the inductor provides an opportunity to enhance the overall inductance by adjusting the material properties of the medium in which the magnetic field is present. The immersion of a planar coil in a material with a high permeability can greatly increase the inductance of the coil. This leads to a reduction in the layout area necessary for a given value of inductance [1-5].

The ability to integrate a material with a high permeability on chip, allows for magnetically coupled circuits and structures to be designed and incorporated along side CMOS circuitry. Devices ranging from A.C. transformers to magnetically driven MEMS structures can be designed and fabricated. An additional application is the incorporation of high frequency microwave ferrites on chip. This allows for the combination of devices such as circulators and other non-reciprocal microwave networks along with silicon digital circuitry [6].

Manganese-Zinc (Mn-Zn) and Nickel-Zinc (Ni-Zn) soft ferrites exhibit good magnetic properties and can be used at high frequencies without laminating. Mn-Zn ferrites exhibit higher permeability and saturation magnetization and are suitable up to 3 MHz. Ni-Zn ferrites have a very high receptivity and are most suitable for frequencies over 1 MHz because of its higher resistivity. Major applications of ferrites are inverter transformers, current and pulse transformers, fly-back and driver transformers, line filters, choke, noise suppressors, magnetic heads and electromagnetic wave absorbers [7, 8].

Electrophoresis (EP) is a process by which charged particles suspended in a solution are made to migrate by an electric field and are deposited on an electrode. One of the first EP coating applications was in forming an insulating coating of alumina onto tungsten filaments. Subsequently, EP coatings have been utilized in the fabrication of phosphor screens for high-resolution cathode ray tubes (CRT) displays [9-11]. In this study, the electrophoretic solution bath is composed of isopropyl alcohol with traces of $Mg(NO_3)_2$ and $La(NO_3)_3$ salts. Glycerol is added to the solution bath as a surfactant to promote increased substrate adhesion. The dissociation of magnesium nitrate in the solution bath charges the ferrite particles. An electric field of ~ 50-160 V/cm is applied with negative terminal connected to the wafer to be plated and aluminum electrode is used as the anode. The substrate is made conducting where the coating is desired.

EXPERIMENTAL DETAILS

Ferrite core inductors have been integrated with aluminum gate MOS transistors and capacitors fabricated using standard semiconductor processing. The process involved eight mask levels that were designed using Mentor Graphics CAD tool and written on 5x5 inch chrome coated quartz plates using MEBES III electron beam writer. The cross section shown in Fig. 1 illustrates the fabrication process. The entire process comprises of three major parts- 1) MOS structures fabrication; 2) copper plating and chemical mechanical planarization; and 3) electrophoretic deposition of ferrite.

An n-well PMOS process was utilized using 4 inch diameter, p type (100) silicon wafers with a resistivity range of 10-25 ohm cm. The first level of metal shown in Fig. 1 is aluminum which makes contact to the transistor active areas. A thick film of SiO_2 is deposited using plasma enhanced chemical vapor deposition (PECVD). Subsequently, patterning of the inductor and interconnect is performed. Via connections are defined with a next lithography step, and etched to produce the cross section seen in Fig.1. A layer of tantalum (50nm) and a layer of copper (100nm) are sputter deposited to provide adhesion and a seed respectively for electroplating of thick copper. Copper is then electroplated and polished back using chemical mechanical planarization (CMP). For electro-chemical deposition of copper, a bath containing of 0.2M $CuSO_4$ (cupric sulfate) and 0.2M H_2SO_4 was used. The total volume of the circulated solution was about 15 liter. The solution contained also 0.001M acetonitrile as a cuprous-complexing ligand to induce bottom up filling. A K-grove pad with an optimal slurry (EKC 9000) and copper removal with a downward force of 5 psi at 90 Hz frequency between the platen and table were used for CMP. This recipe was followed by a barrier removal oxidizer for tantalum and was assisted mechanically using 90 Hz frequency and 3 psi down force with both slurry flow rates at 100 sccm.

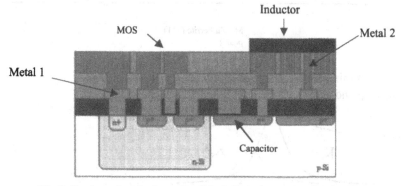

Fig.1. Cross-sectional schematic of the inductor and capacitor integrated MOS

To demonstrate the integration of electrophoretic deposition of ferrite material, a soft ceramic ferrite – Mn-Zn ferrite was chosen for this study. The material used for this study was acquired from a commercial source – Ferronics, Inc. in the form of sintered toroidal cores. The initial permeability of this material is ~ 15,000 [12]. The toroids were ball milled with ½ inch zirconium balls in a zirconium oxide container using a two-clamp laboratory mill (8000D SPEX CentriPrep Dual Mixer/Mill) to create a sub-micron powder. This particle distribution was estimated using Sherrer analysis of the x-ray diffraction pattern of the powders acquired using the Rigaku DMAX-IIB powder diffractometer.

RESULTS

The ground ferrite was characterized for the particle size distribution. The (311) ferrite peak was chosen for calculating the particle size. The Pearson type VII function was assumed to describe the peak. Fig. 2 shows the (311) peak of the Mn-Zn x-ray diffraction pattern for 5 minutes and 60 minutes grinding processes. On deconvulation, $K_{\alpha 1}$ peak was obtained which was analyzed for coefficient b, proportional to the FWHM and the shape factor m. The measured FWHM can be expressed as [13]

$$B_{meas} = FWHM = 2b\sqrt{\left(2^{1/m} - 1\right)} \tag{1}$$

The full width half maxima (FWHM) due to the particle size, $B_{particle+strain}$ can be obtained from the measured FWHM B_{meas}.

$$B^2_{particle+strain} = B^2_{meas} - B^2_{inst} \tag{2}$$

Assuming that the FWHM (B_{inst}) does not change with grinding time, $B_{particle+strain}$ can be obtained. The average particle size is then obtained using the Scherrer's formula [14]

$$x = \frac{0.9\lambda}{B_{particle} \cos \theta} \tag{3}$$

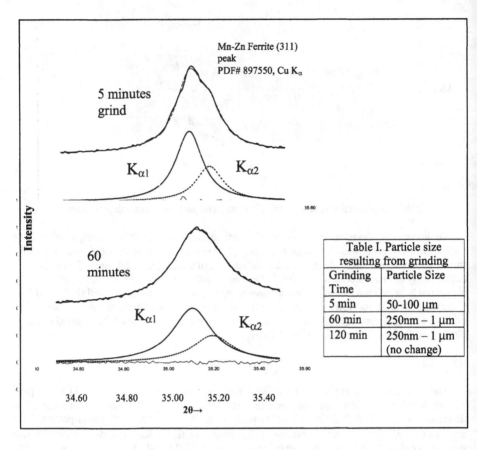

Fig. 2. (311) XRD peak of ferrite powder after 5 and 60 minutes of grinding.

In this analysis, if broadening due to strain is assumed to be negligible an average particle size was estimated to be ~ 50 nm. The scanning electron microscope measurements give a particle size range as given in the Table I. These results show that microstrains may have been caused as a result of the grinding process. In addition, agglomeration of ceramic nano-particles produces an increase in the visible particle distribution.

Fig. 3 and Fig. 4 show the effect of the magnesium nitrate concentration and the voltage bias respectively on the current which determines the deposition rate. The three curves shown in the Fig. 3 correspond to 0.5 g ,1.0 g and 1.5g of magnesium nitrate dissolved in 100ml of IPA keeping lanthanum nitrate constant at 1 g. It can be observed that most of the deposition occurs during the first 10 minutes and the process is self limiting and there is an optimum concentration of magnesium salt for obtaining a higher deposition rate.

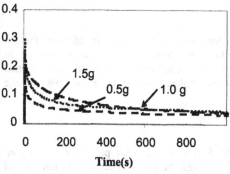

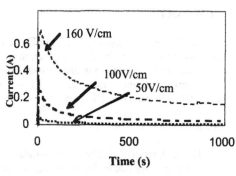

1.5g 0.5g 1.0 g

Time(s)

Fig. 3. Electrophoretic current as a function of time for different amounts of Mg(NO₃)₂

160 V/cm

100V/cm

50V/cm

Time (s)

Fig. 4. Electrophoretic current as a function of time for different applied electric field.

Fig. 5 shows the magnetization curve of the film obtained by vibrating sample magnetometer (VSM). A scanning micrographs of the ferrite included in the Fig. 5 shows selective deposition on the copper inductor coil (top left). The electrical measurements made on the inductors show a moderate enhancement in the inductance value indicating a much lower initial permeability of the powder material as compared to the original toroidal material as shown by the *weak* magnetization curve obtained. Further investigations are necessary to improve the packing density and the magnetic properties of the powder material.

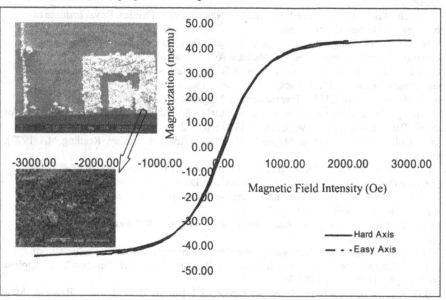

Fig. 5. Magnetization curve of the ferrite film deposited by electrophoretic process having a microstructure shown on the lower left.

SUMMARY

Electrophoretic deposition of Mn-Zn ferrite has been demonstrated. In addition, the EP process has been employed in conjunction with chemical mechanical planarization of copper to fabricate ferrite core micro inductors integrated with MOS circuitry. This method can be applied to microwave magnetic materials for RF and wireless applications if suitable materials be made available with nano particle sizes.

ACKNOWLEDGMENTS

The authors would like to acknowledge Dr. Surendra Gupta for particle size analysis. Special thanks are due to Ferronics Inc., for providing with the ferrite material and Chih Ling of Veeco for providing the VSM data. The work has been supported by the National Science Foundation for under the grant No. ECS-0219379.

REFERENCES

[1] M. Yamaguchi "Characteristics and Analysis of a thin film inductor with closed magnetic circuit structure" IEEE Transaction on Magnetics, Vol.28, No.5, September 1992
[2] M. Yamaguchi " Sandwich type Ferromagnetic RF integrated inductor" IEEE Transaction on Microwave Theory and Techniques Vol.49, No.12, December 2001
[3] S. Sim " High Frequency on chip inductance Model" IEEE Electron Device Letters, Vol.23, No.12, December 2002
[4] V. Korenivski " GHz magnetic film inductor" Nanostructure Physics, Royal Institute of Technology, 10044 Stockholm Sweden
[5] R.B. Van Dover, "Magnetic film inductor for radio frequency applications", Bell Laboratories, Lucent Technologies, 700 Moutain Ave. Muray Hill, NY
[6] C.Ahn " Micromachined Planar inductors on Silicon Wafers for MEMS applications", IEEE Transactions on Industrial Electronics, Vol. 45, No. 6 December 1998
based on Magnetic Films", IEEE Transactions on Magnetic, Vol. 34, No.4, July 1998
[7] J. Baszynski, in *Ferrites* edited by H. Watanabe, S. Lida and M. Sugimoto, (Center for Academic Publications Japan, Kyoto, Jaspan 1980) p. 212
[8] B.D. Cullity, *Introduction to Magnetic Materials*, (Addison Wesley, Reading MA 1972), Chapter 6.
[9] J.Talbot " Electrophoretic Deposition in the Processing of Information Display's" , Chemical Engineering Department, University of California, San Diego 9500 Gilman Drive, CA 92093
[10] J.Talbot, D.Russ " A Method for measuring the Adhesion strength of Powder Coatings" , Journal of Adhesion, 1998, Vol.68, pp. 257-268
[11] S. Kurinec and E. Sluzky *"Method of Fabricating an Ultra High Resolution Three Color Screen"*, Pat. No.5,582,703, Dec 1996.
[12] Ferronics V Material product catalog: http://www.ferronics.com/catalog/materials.pdf
[13] S. K. Gupta, "Peak Decomposition using Pearson Type VII Function", J. Applied Crystallography, (1998), 31, 474-476.
[14] B.D. Cullity, Elements of X-ray Diffraction, 2nd Ed. (Addison-Wesley, Reading, MA, 1978), pp. 101-102.

Making Wafer Bonding Viable for Mass Production

Cher Ming Tan*, Wei Bo Yu, and Wei Jun[1]
School of EEE, Nanyang Technological University,
Block S2, Nanyang Avenue, Singapore 639798
*E-mail: ecmtan@ntu.edu.sg
[1]Singapore Institute of Manufacturing Technology, Singapore 638075

ABSTRACT

Direct wafer bonding was performed under medium vacuum condition. High bonding strength (larger than 20 MPa) is achieved at the bonding temperature of only 400°C, and the annealing time for complete bonding is less than 5 hours. The bonding efficiency (percentage of the bonded area over entire wafer area) of the medium vacuum wafer bonding (MVWB) is also found to be better than the traditional wafer bonding.

Qualitative description of the mechanism of MVWB is proposed in present work. It is believed that the medium vacuum can enhance the out-diffusion of the water molecules and other trapped impurities through the bonding interface which is porous initially. This enhanced diffusion speeds up the chemical reaction for the formation of Si-O-Si, and thus more bonding sites are available before the interface close-up. As a result, we observe an increase in bonding strength, bonding efficiency and the bonding speed.

INTRODUCTION

Wafer bonding is an attractive wafer fabrication technology for System on Chip (SoC), Micro-Electro-Mechanics-System (MEMS), Silicon on Insulator (SOI) devices etc. Traditional direct wafer bonding usually requires high temperature annealing above 800-1000 °C to achieve strong bonding. However, this high temperature treatment will generate several problems, such as thermally induced mechanical stress due to the difference in thermal expansion coefficients of the materials, undesirable changes and reactions for the materials and structures that are sensitive to high temperatures, and so on. Therefore, it is necessary to find a bonding process that can result in strong bonding at temperatures as low as possible (e.g. below 400 °C).

There are many methods to achieve high bonding strength at low temperature, such as plasma activated wafer bonding [1], hot nitric acid dipping wafer bonding [2] and vacuum wafer bonding [3-7]. Investigations showed that vacuum wafer bonding has the potential of becoming a reliable low temperature bonding method. These investigations include wafer bonding in ultrahigh vacuum (UHV) (10^{-10}mbar) [3], wafer bonding after Ar beam surface activation in high vacuum (10^{-8}mbar) [4], wafer bonding at medium vacuum level (10^{-4}mbar) [5,6], and wafer bonding at low vacuum level (several mbar) [7].

The effect of bonding parameters on the bonding quality has been studied extensively for the traditional wafer bonding. The bonding quality studied are the bonding efficiency, i.e. the ratio of the bonded area to the entire wafer area, and the bonding strength. If wafer bonding is to be employed for mass production, the bonding speed is an important parameter. For traditional wafer bonding, it is found that more than one hundred hours is required to achieve saturation of the bonding strength, which is apparently not economical.

As low temperature vacuum wafer bonding is promising in term of its bonding quality, we would like to study the bonding speed for MVWB in this work. The effect of annealing temperature and time on the bonding efficiency, the bonding strength and the bonding speed will be investigated.

EXPERIMENTAL DETAILS

Both the traditional direct wafer bonding and medium vacuum wafer bonding were performed in this work as shown in Table 1 for comparison purpose. The silicon wafers used are 4", 500 μm thick p-type, (100) standard bare wafers with resistivity of 1-50 Ω-cm. The oxide wafers are silicon wafers with thermal oxide of 500-600 nm thickness, and the total thickness is 500 μm. The RCA cleaning of wafers involves 10 minutes of RCA1 ($NH_4OH:H_2O_2:H_2O = 0.05:1:5$) cleaning at 70 °C [8], 5 minutes of deionized (DI) water rinsing and argon gas drying. Traditional and vacuum wafer bonding were done in oven (Latent) and vacuum wafer bonder (EV501, EVG) respectively. All these processes were performed in a class 10k clean room environment.

Table 1 Different wafer bonding processes studied in this work

Wafer no.		RCA Cleaning	Wafer Contacting	Process		
				Annealing		
				Atmosphere	°C	Time (hour)
Vacuum Bonding	A1	X*	X	V**	200	1-5
	A2	X	X	V	300	1-5
	A3	X	X	V	400	1-5
	A4	X	X	V	500	1-5
Traditional Bonding	B1	X	X	A***	200	1-5
	B2	X	X	A	300	1-5
	B3	X	X	A	400	1-5
	B4	X	X	A	500	1-5

*X: means the process step was carried out; **V: Vacuum (10^{-4}mbar); ***A: Air

After drying, the two as-bonded wafers were contacted with each other in air from one edge as shown in Figure 1. In this way, most of the air could be squeezed out and less bubble will be produced.

For vacuum wafer bonding, group A processes were designed. In these processes, Si and SiO_2 wafers were loaded into the vacuum bonder after being contacted in the air. Vacuum was then applied. When the vacuum level reached 10^{-4} mbar, heating began in such a rate that the temperature reaches the preset annealing temperature (200, 300, 400, 500 °C respectively) in 30 minutes and then maintained at the temperature for different times (1, 2, 3, 4, 5 hours). Thereafter, the bonded pair was cooled down to room temperature in 2 hours. The vacuum level was maintained throughout the entire bonding process.

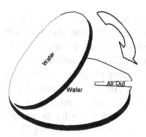

Figure 1 Schematic of room temperature wafer contact

The processes of traditional wafer bonding are labeled as group B shown in Table 1. Two wafers were bonded in the air at room temperature and then loaded into oven. The oven began to heat up to the preset annealing temperature (200, 300, 400, 500 °C respectively) and maintain the temperature for different times (1, 2, 3, 4, 5 hours). Finally, the oven was switched off and cooled to room temperature.

After bonding, the bonded pairs were examined under scanning acoustic microscopy (SAM) using C-SAM SERIES D6000 from SONOSCAN Inc. The resolution of the SAM is 2.5 μm. MATLAB 6.5 was employed to analyze the SAM micrographs to calculate the percentage of the unbonded area.

To measure the bonding strength, pull test was chosen. The bonded pairs were diced into 10×10 mm² pieces for bonding strength measurement with an Instron tensile testing machine (model 4505). Prior to the tensile test, the top and bottom surfaces of the bonded pairs were roughened with abrasive paper. Ultrasonic cleaning in acetone of the bonded pair was then performed to ensure no fracture occur at these two joint interfaces (interfaces between steel pad and bonded pair). The bonded pairs were then glued onto two steel pads that were attached to the pull test machine using a room temperature cured epoxy adhesive. For each bonding condition, six samples were tested and the bonding strength was calculated based on the average fracture force.

RESULTS

The results of bonding strength within 5 hours of annealing using the processes in groups A and B are shown in Figure 2. From this figure, three phenomena can be observed:

a) In both bonding methods, the bonding strength gradually increases with annealing temperature from 200 °C to 500 °C. This is consistent with the trend of the dependence of bonding strength on the annealing temperature [9].

b) It is apparent that the medium vacuum bonding can enhance the bonding strength compared to the traditional bonding at all annealing temperatures for the short annealing time used in this work. In all cases of vacuum wafer bonding, the bonding strengths are above 7 MPa, which is high enough for dicing. Also, it can be observed that the bonding strength after 300 °C annealing increases abruptly to above 20 MPa.

c) The saturated bonding strength can be achieved in much shorter time. Our experimental results showed that the bonding strength does not improved after 5 hours under medium vacuum as compared to that of tradition wafer bonding which always takes more than hundred hours [10]. In other words, vacuum can accelerate the bonding process. This shows the potential of medium vacuum wafer bonding for mass-production applications, alone or combined with other surface activation methods.

The three important phenomena observed could be interpreted by the mechanism of medium vacuum wafer bonding. Although the model proposed by Stengl [9] and Tong [10] well explains the traditional wafer bonding, it is not sufficient to explain the enhancement of bonding strength in the low temperature wafer bonding, such as plasma activated wafer bonding and vacuum wafer bonding [1,7]. In fact, for the vacuum wafer bonding, the bonding mechanisms at different level of vacuum are found to be distinct. For low vacuum level wafer bonding, the enhancement of the bonding strength is proposed to be due to the reduction of trapped gas at the bonding interface, which in turn increases the bondable area at the bonding interface and allow for the development of covalent bonds (siloxane) [7]. However, in UHV wafer bonding, two atomically clean surfaces produced in UHV make covalent bonding easier, thus attaining high bonding strength [3]. In this work, because the two wafers were pre-bonded at room temperature prior to loading into the vacuum chamber in our experiments, medium vacuum cannot have the effect as UHV does.[3] Therefore, the mechanism of UHV wafer bonding is not applicable in our case.

The mechanism of the medium vacuum level wafer bonding is rarely discussed [5]. However, from the above discussion, it is reasonable to propose that the medium vacuum produces a similar but stronger effect as low vacuum level during the wafer bonding, i.e. the medium vacuum accelerates the diffusion of trapped impurities, such as water, nitrogen and hydrogen [11]. Furthermore, the bonding interface can be considered as porous in nature immediately after room temperature bonding due to the surface micro-roughness and uncovered area by the attached water molecules [10]. As more bonding sites are bonded through Equation (1), the interface gradually close up, and thus changing the interlayer to amorphous like SiO_2 in the annealing process.

$$Si-OH+HO-Si \leftrightarrow Si-O-Si+HOH \qquad (1)$$

Using the proposed bonding mechanism, one can see that the medium vacuum can enhance the out-diffusion of the water molecules and other trapped impurities [11] at the initial porous interface, hence speed up the reaction (1) and the formation of Si-O-Si., and thus more bonding sites can be bonded before the interface close-up. This results in an increase in bonding strength and bonding efficiency as well as the bonding speed (i.e. the time taken for the bonding strength to reach saturation value is shorter). At higher temperature, the diffusion coefficients of water and other trapped impurities should be higher. This can explain the dependence of bonding strength on annealing temperature.

Based on the above understanding, mathematical model was developed [12]. It shows that the good agreement between the model and the experimental data is observed.

To show the bonding efficiency of the medium vacuum wafer bonding, the SAM results of the bonded wafers under 2 hours of annealing at different temperatures of 200, 300, 400 and 500°C respectively are shown in Figure 3. Only one representative SAM picture for each annealing process is shown. One can see that the unbonded area of

MVWB is decreased compared with the traditional bonding, and it is no longer a serious problem when annealing temperature is higher than 300 °C.

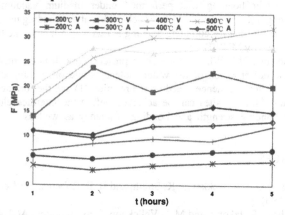

Figure 2 Experiments results of bonding strength after different bonding processes, e.g., different annealing temperature, annealing time and annealing environment (vacuum and air atmosphere). "V": denotes vacuum wafer bonding; "A": denotes traditional air wafer bonding.

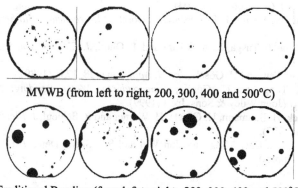

MVWB (from left to right, 200, 300, 400 and 500°C)

Traditional Bonding (from left to right , 200, 300, 400 and 500°C)

Figure 3 Comparison of the bonding efficiency between MVWB and traditional wafer bonding

CONCLUSIONS

Direct wafer bonding was performed under medium vacuum condition. High bonding strength (larger than 20 MPa) is achievable at the bonding temperature of 400°C, and the annealing time for complete bonding is less than 5 hours. The bonding efficiency is also improved by MVWB.

Mechanism of MVWB is proposed in present work. It is the medium vacuum that enhances the out-diffusion of the water molecules and other trapped impurities at the initial porous interface, hence speed up the reaction (1) and the formation of Si-O-Si. As a result, more bonding sites can be achieved before the interface close-up. Hence, an increase in bonding strength and bonding efficiency as well as the bonding speed is observed.

REFERENCES

[1] S. N. Farrens, J. R. Dekker, J. K. Smith, and B. E. Roberds, *J. Electrochem. Soc.* **142**, 3949 (1995).

[2] A. Berthold, B. Jakoby, and M. J. Vellekoop, *Sens. Actuators* **A68**, 410 (1998).

[3] U. Gösele, H. Stenzel, T. Martini, J. Steinkirchner, D. Conrad, and K. Scheerschmidt, *Appl. Phys. Lett.* **67**, 3614 (1995).

[4] H. Takagi, K. Kikuchi, and R. Maeda, *Jpn. J. Appl. Phys.* **37**, 4197 (1998).

[5] W. B. Yu, C. M. Tan, J. Wei, S. S. Deng and S. M. L. Nai, *Sens. Actuators* **A115**, 67 (2004)

[6] W. B. Yu, C. M. Tan, J. Wei, S. S. Deng and S. M. L. Nai, *IEEE Proc. 5th Electron. Packag. Tech. Conf.*, pp. 294 (2003).

[7] Q. Y. Tong, W. J. Kim, T. H. Lee, and U. Gösele, *Electrochem. Solid-State Lett.* **1**, 52 (1998).

[8] M. Miyashita, T. Tusga, K. Makihara and T. Ohmi, *J. Electrochem. Soc.* **139**, 2133 (1992).

[9] R. Stengl, T. Tan, and U. Gösele, *Jpn. J. Appl. Phys.* **28**, 1735 (1989).

[10] Q. Y. Tong and U. Gösele, *Semiconductor Wafer Bonding: Science and Technology* (John Wiley & Sons, Inc.) (1999).

[11] S. Mack, H. Baumann, U. Gösele, H. Werner and R. Schlögl, *J. Electrochem. Soc.* **144**, 1106 (1997).

[12] Wei Bo Yu, Cher Ming Tan, Jun Wei and Shu Sheng Deng, *IEEE Trans. Adv. Packag.* (in press).

AUTHOR INDEX

SUBJECT INDEX

amorphous silicon, 3, 15, 39
a-Si:H, 21

barium strontium titanate, 133

chemical sensors, 103
CMOS(-), 21, 91
    integrated, 109
    technology, 53
cofiring, 151
color, 39

dark current, 15, 27
deposition, 133
detector, 65
device physics, 21
dieletric, 151
DNA microarray, 91

electro-luminescence, 83
electrophoretic, 157

GaAs, 83
gallium arsenide, 77
germanium, 77
gold, 91

hafnium oxide, 145
hollow cathode, 133

integrated, 65
ITO, 139

lab-on-microchip, 119
liquid chemical sensor, 109
low temperature, 163
LTCC, 151

metal induced crystallization, 45

micro total analysis system, 119
microfluidics, 103
micro-inductors, 157
microwave focusing, 71
Mn-Zn ferrite, 157
monolithic instrument(s), 53, 119

nickel, 45
noise, 27

OLED, 139
optical elements based on photonic
    crystals, 71

photonic crystals, 71
plastic substrate, 139
PLD, 127
polycrystalline silicon, 45
polyimide, 127
porous silicon, 103

sensor(s), 3, 39
silicon, 65, 77, 163
sputtered, 145
structure, 145
subwavelength photonics, 53
surface acoustic wave, 83

TFA
    sensor, 27
    technology, 3
thin-film fabrication, 109

vertical integration, 15

wafer bonding, 163

zinc oxide, 127

Printed in the United States
By Bookmasters